MILK!

FICTION

City Beasts: Fourteen Stories of Uninvited Wildlife

Edible Stories: A Novel in Sixteen Parts

*"The Belly of Paris" by Émile Zola: A New Translation with an Introduction
by Mark Kurlansky*

Boogaloo on 2nd Avenue: A Novel of Pastry, Guilt, and Music

The White Man in the Tree and Other Stories

CHILDREN/YOUNG ADULT

Frozen in Time: Clarence Birdseye's Outrageous Idea about Frozen Food

Battle Fatigue

World without Fish

The Story of Salt

The Girl Who Swam to Euskadi

The Cod's Tale

—A 10,000-YEAR FOOD FRACAS—

MARK KURLANSKY

BLOOMSBURY PUBLISHING

NEW YORK · LONDON · OXFORD · NEW DELHI · SYDNEY

BLOOMSBURY PUBLISHING
Bloomsbury Publishing Inc.
1385 Broadway, New York, NY 10018, USA

BLOOMSBURY, BLOOMSBURY PUBLISHING, and the Diana logo are
trademarks of Bloomsbury Publishing Plc

First published in the United States 2018

Copyright © Mark Kurlansky, 2018
Illustrations on pages 8, 17, 58, 99, 195, 196, 211, 212, 232, 244, 257, 270, 274, 279, 285,
297, 298, 306, 322 © Mark Kurlansky, 2018

ISBN: HB: 978-1-63286-382-9; eBook: 978-1-63286-384-3

LIBRARY OF CONGRESS CATALOGING-IN-PUBLICATION DATA

Names: Kurlansky, Mark, author.
Title: Milk! : a 10,000-year food fracas / Mark Kurlansky.
Description: New York : Bloomsbury, 2018. | Includes bibliographical references
and index.
Identifiers: LCCN 2017039795 | ISBN 9781632863829 (hardcover) |
ISBN 9781632863843 (ebook)
Subjects: LCSH: Dairy products—History. | Milk—History.
Classification: LCC SF250.5 .K87 2018 | DDC 637/.109—dc23 LC record available at
https://lccn.loc.gov/2017039795

2 4 6 8 10 9 7 5 3 1

Designed by Simon Sullivan
Typeset by Westchester Publishing Services
Printed and bound in the U.S.A. by Berryville Graphics Inc., Berryville, Virginia

To find out more about our authors and books visit www.bloomsbury.com and
sign up for our newsletters.

Bloomsbury books may be purchased for business or promotional use. For
information on bulk purchases please contact Macmillan Corporate and Premium
Sales Department at specialmarkets@macmillan.com.

To my dear friend Charlotte Sheedy,
one of the finest people I have ever known

How sad that we have no memories of our mother's milk or our first sight of the world, through eyes made blurry by the tears we shed for milk . . .

—Sait Faik Abasiyanik, "Milk"

CONTENTS

PART THREE **COWS AND TRUTH**

A NOTE ABOUT THE RECIPES

*Si vous n'êtes pas capable d'un peu d'alchémie, ce n'est pas la
peine de vous mettre en cuisine.*
If you are not capable of a bit of alchemy, don't bother
going into the kitchen.

—COLETTE

OFTEN INCLUDE RECIPES in my books, not because of some suppressed
desire to write cookbooks, but because I believe recipes to be invalu-
able artifacts. They teach us about societies and the social order in
which they were created. They tell us what life was like at the time
when those dishes were first cooked. I usually have not concerned
myself with whether the recipe will result in a pleasing dish.

However, in researching this book, I came across so many dairy
dishes to choose from that I decided to select ones that would
be enjoyable to eat. And so I encourage the reader to try some of
these recipes. That said, I would avoid Richard Nixon's cottage
cheese recipe, "milk water," or some of the baby formulas. Milk toast
never appealed to me, either. But cream pancakes and junket, syllabub
and posset are voices from the past worth trying. Indian pudding,
ginger ice cream, a hot strawberry sundae, Jamaican banana ice
cream, or especially Pellegrino Artusi's caffè latte gelato could
make your day. Do try Louis Diat's glorious original recipe for
vichyssoise. And there are many others, such as the Indian recipes
from what is possibly the greatest dairy cuisine.

I always included the original recipes without updating them,
though I have slipped in a few explanations [between brackets], as

some of the recipes, particularly the older ones, are a bit vague and hard to follow. One of those is the recipe for Cato's cheesecake, which many have tried to clarify, but no one seems to know exactly how it is supposed to turn out. That means that your version is as valid as the next. Feel free to make adjustments, create, and use modern tools such as a stove and an ice cream maker. The best dishes are often those to which the chef adds a personal statement, even when using a very old recipe. As Colette said, use a little alchemy.

—PART ONE—

THE SAFETY OF CURDS

He curdled half the milk and set it aside in wicker strainers, but the other half he poured into bowls that he might drink it for his supper.

—ODYSSEUS, observing the
Cyclops in Homer's *The Odyssey*

1

THE FIRST TASTE OF SWEETNESS

SINCE MILK IS a food and this book contains 126 recipes, it might seem as if this should be a food book. But milk is a food with a history—it has been argued about for at least the past ten thousand years. It is the most argued-over food in human history, which is why it was the first food to find its way into a modern scientific laboratory and why it is the most regulated of all foods.

People have argued over the importance of breastfeeding, the proper role of mothers, the healthful versus unhealthful qualities of milk, the best sources of milk, farming practices, animal rights, raw versus pasteurized milk, the safety of raw milk cheese, the proper role of government, the organic food movement, hormones, genetically modified crops, and more.

Here is a food fight that gourmets, chefs, agronomists, parents, feminists, chemists, epidemiologists, nutritionists, biologists, economists, and animal lovers can all weigh in on.

One great misconception about milk is that people who cannot drink it have something wrong with them. In truth, the aberrant condition is being able to drink milk. Milk drinkers are mostly of European extraction, and as we are living in a Eurocentric world, we tend to think of consuming dairy products as a normal thing to do—something that is forgone in some regions only because of a malady known as lactose intolerance. But lactose intolerance is the natural condition of all mammals. Humans are the only mammals that consume milk past weaning, apparently in defiance of a basic rule of nature. In nature, the babies of most mammals nurse only until they are ready for food, and then a gene steps in to shut down the ability to digest milk. Lactose, a sugar in milk, is digestible only when lactase, a genetically controlled enzyme, is present in the intestines. Almost everyone is born with lactase. Without it, a baby could not breastfeed. But as most babies get older, a gene cuts off the production of lactase and they can no longer consume milk.

But something went wrong with Europeans—as well as Middle Easterners, North Africans, and people from the Indian subcontinent. They lack the gene and so continue to produce lactase and consume milk into adulthood.

The gene travels in blood-related tribes and family groups. So though most black Africans are lactose-intolerant, the Masai, who are cattle herders, are not. Those who are intolerant tend not to have dairy in their culture. But in societies that do adopt a dairy culture, such as the Masai or Indians in Asia, the ability to digest milk remains. The early Europeans had dairy cultures and so were lactose-tolerant, though this was truer in the north, where short growing seasons necessitated a supplemental food source. However, being lactose-tolerant certainly is not entirely a question of climate, because the original Americans—occupying two continents stretching from Patagonia to Alaska and including just about every imaginable climate—were lactose-intolerant.

Though most Europeans drink milk today, we don't really know the original extent of lactose intolerance on the Continent, because centuries ago, milk drinking there was rare. Hard cheese and yogurt were popular, but they do not contain lactose, and this might have been a reason why the Europeans favored them. Somewhere between then and now, though, Europeans began to drink milk, and because they always had a way of defining their aberrations as the norm, they took their dairy animals with them wherever they went around the globe.

★   ★   ★

To THINK OF milk as just another food would be to ignore the galaxy we live in. Not just figuratively, but also literally. Our galaxy is called the Milky Way, and both it and the word "galaxy" have their origins in the Greek word for milk, *gala*. According to Greek mythology, the Milky Way was formed when Hera, the Greek goddess of womanhood, spilled milk while breastfeeding Heracles, known to the Romans as Hercules. Each drop became a speck of light, known to us as a star. And Hera must have spilled a lot of milk, because modern astronomers estimate that there are 400 billion stars in the galaxy.

Numerous cultures have such milk-based creation myths. The Fulani people of West Africa believe that the world started with a huge drop of milk from which everything else was created. According to Norse legend, in the beginning there was a giant frost ogre named Ymir, who was sustained by a cow made from thawing frost. From her four teats ran four rivers of milk that fed the emerging world.

In what is today Iraq, the Sumerian culture, the first civilization to develop a written language, was among the first to milk domesticated animals. According to one of their legends, a priest named Shamash in the city of Urak spoke to animals and persuaded them to withhold their milk from the goddess Nidaba. But two shepherd brothers, discovering the plot, threw Shamash into the Euphrates, where he transformed himself into a sheep. The brothers discovered his ruse and threw him into the Euphrates again. This time, he turned himself into a cow. Discovered a third time, he assumed the form of a chamois, a kind of antelope. This appears to be a legend about a search for a reliable milking animal.

Isis, the Egyptian goddess of motherhood, giver of life, was often shown breastfeeding a pharaoh, while Osiris, her husband, was celebrated for pouring out bowls of milk, one for each day of the year. Isis was a popular deity throughout the Middle East, and was depicted with large breasts and a cow's head and horns. Images of her Greek counterpart, Artemis, sometimes had several dozen breasts. The

Egyptians also worshiped another deity, Hathor, as the cow goddess. Milk was a common offering in Egyptian temples.

It was believed that a baby acquired the personality of his or her wet nurse, so these caregivers needed to be carefully selected. It was said that Zeus was so unfaithful to women because he had been suckled by a goat, an animal infamous for debauchery, on the island of Crete. Infants who were suckled by the same wet nurse were regarded as milk siblings and were forbidden by the Assyrians from intermarrying.

In the third or second century B.C.E., a letter written to a new Roman mother stated: "The wet nurse should not be temperamental or talkative nor uncontrolled in her appetite for food, but orderly and temperate, practical, not a foreigner, but a Greek." This last requirement came up frequently in ancient Greece. Soranus, the first- and second-century A.D. Greek physician, repeatedly told his Greco-Roman audience that wet nurses should be Greek.

Hindus revered, and continue to revere, cows. In Sanskrit the word for cow is *aghnya*, which means "that which cannot be slaughtered." Hinduism has a creation myth in which the god Vishnu churns a sea of milk to create the universe.

The early Christians regarded all this cow worship as pagan, but still kept a special place for milk—human milk, that is—in their religion. The Virgin Mary was continually depicted exposing a breast and lactating. A leading Christian figure, the twelfth-century Bernard of Clairvaux, was said to have derived his inspiration from the Virgin Mary, who had appeared to him, bared a breast, and squirted three drops of milk into his mouth.

Medieval Christianity abounds with stories of people drinking Mary's milk, and even, in a few inexplicable cases, Christ's milk. These people received not Bernard's circumspect three drops, but rather a long, arched stream, at least according to some artists. One ignorant monk was said to have acquired great wisdom when the soft, sweet-spoken Mary bade him come close, bared her breasts, and had him suck at length on them. All this reflects the old and enduring

THE FIRST TASTE OF SWEETNESS · 7

Christian belief that the breastfed child acquired the traits of the woman who fed her or him.

Christians in the Middle Ages thought that milk was blood that had turned white when it traveled to the breast, which is why milk was banned on meatless holy days—more than half the days of the year. Japanese Buddhists had the same belief and avoided consuming dairy products. They looked down on Westerners, who they thought consumed too much dairy. They claimed they could smell it on them, and even into the twentieth century used the pejorative term *Batā dasaku*, "butter stinker," for a Westerner.

Nor have Jews ever been comfortable with the consumption of dairy. In Exodus, it states, "You shall not boil a young goat in its mother's milk." This has been interpreted as an absolute ban on eating any meat product, even chicken, in the same meal as any dairy product.

And yet even in ancient times, there have always been those who insisted on the health-giving qualities of milk. A Sumerian cuneiform tablet states that milk and *laban*, a yogurt-like sour drink, drive off illness. Pliny the Elder, the first-century A.D. Roman writer, claimed that milk was an effective antidote for those who swallowed quicksilver.

PRODUCING MILK IS what defines a mammal. The scientific class Mammalia, to which humans belong, is so labeled from the Latin *mammal*, meaning "of the breast." We are the milk-producing class of animals, and we share milk among ourselves, although animals other than humans usually drink only the milk of their mothers unless humans intervene.

Still, to varying degrees, most mammal milk is acceptable food. And which milk is the best milk is one of history's unending debates. There is not even universal agreement that human milk is best for humans.

Different milks have varying amounts of fats, proteins, and lactose, and arguments abound over the relative merits and possible harms

Jersey

of each. For many centuries, milk with a high fat content was considered to be the best, and low-fat or skimmed milk was considered fraudulent—in fact, selling it was often illegal. Cow breeds that produce milk with a high fat content, such as Ayrshires, Jerseys, and Guernseys, have always been valued, especially for cheesemaking.

It is often stated that milk, produced by nature to feed newborns, is the ideal food. And it has long been recognized that newborn humans, calves, and lambs do not have the exact same needs. Early on, people understood that the milks of different species varied somehow, though it wasn't until the eighteenth century that those differences were quantified.

Each species has its own unique milk, designed by nature to meet its needs. Young whales have to build a layer of fat quickly in order to survive, and so whale milk is 34.8 percent fat, as opposed to human milk, which is only 4.5 percent fat. Northern seals also have to acquire fat quickly; a gray seal's milk is 53.2 percent fat, making it about the fattiest milk of all. Even aside from the logistical problems of milking gray seals, let alone whales, their high-fat milk is not suitable for us.

Human babies like and need milk that is 4.5 percent fat, only 1.1 percent protein, 6.8 percent lactose, and about 87 percent water. Not surprisingly, one of the milks closest to that of humans comes from monkeys. But while we have managed, with considerable ease, to accept the idea of feeding our young on the milks of other animals, we prefer that those animals not be too biologically similar. Most societies would find the idea of monkey dairies upsetting.

Milk also contains more than just fat, protein, lactose, and water. There are such things as cholesterol and linoleic acid to consider. For example, buffalo's milk, which people drink in India and the Philippines and use to make mozzarella cheese in southern Italy, has more fat but less cholesterol than cow's milk. Cow's milk also lacks linoleic acid, which is considered important for human brain growth. And while cow's milk has four times as much protein as human milk, most of the extra protein is in the form of casein, which has valuable commercial applications but is not needed in that quantity for a human baby's development.

Human milk has far more lactose than most other milks. Lactose is a sugar, so all milks are a bit sweet, but human milk is particularly sweet. Humans and most other mammals develop a fondness for sweets, probably because of their first food.

Before sugar cane and beet sugar became ubiquitous, honey was the number one sweet available. Milk was a close second, though, which is why the two were often grouped together. In a Vedic song of India, at least as old as the Old Testament, Rigveda says to Indra,

> With honey of the bees is milk mixed,
> Come quick. Run and drink.

In the Old Testament, there are twenty references to milk and honey amid some fifty references to milk alone—human, cow, and goat. Most famously, the Hebrews are promised a land of milk and honey, a land of rare sweets.

Of course, there may have been a purely gastronomic aspect to this. Milk and honey are a pleasing combination. Honey mixed with yogurt, the sweet and the sour, is especially tasty. There may even be a medical reason for the combination.

★   ★   ★

THROUGHOUT HISTORY IT has often been argued that, after human milk, goat's and donkey's milks were most suitable for us because their compositions are closest to ours. But that is not entirely true. Donkey's milk has far less fat than human milk, and goat's milk has triple the amount of protein.

Cows, sheep, goats, and buffalos have four stomachs; camels and llamas have three. Animals that have more than one stomach are known as ruminants. Some ruminants, such as cows and sheep, are grazers who munch on grass, and some, such as goats and deer, instead nibble on nutritious shrubs in the woods.

The word "ruminant" comes from the Latin word *ruminare*, which means "to rechew." Food is regurgitated, rechewed, and sent to the rumen, one of the animal's stomachs, to be decomposed by fermentation before passing on to the other compartments. A cow chews

Etching by Jean-Louis Demarne, 1752–1829. A cow and calf with goats in background. (Author's collection)

for between six and eight hours a day, which produces some 42 gallons of saliva that buffer the acids produced in fermentation.

Animals that have one stomach are known as monogastrics, and it would seem to make sense that milk produced by an animal that digests the way we do would be most suitable for us. This is why even today, donkey's milk is produced commercially, especially in Italy, and sold as a health product.

Another monogastric animal is the horse, but mare's milk has caught on in only a few cultures, perhaps because it is extremely low in fat. Pliny the Elder reported that the Sarmatians, nomadic tribesmen in Iran and the southern Urals, consumed mare's milk mixed with millet, creating a kind of porridge that would become popular in other cultures when millet was mixed with different milks.

Herodotus, the fifth-century B.C.E. Greek historian, wrote that the Scythians, also Eurasian nomads, had a diet that consisted almost entirely of mare's milk. But when Marco Polo, who is credited with introducing many European food trends, reported that the Mongols drank mare's milk, Europeans were not tempted to take up the practice.

And why has the pig, another monogastric animal and the most ubiquitous farm animal in the world, never been called to dairy duty? Perhaps it is because we don't like to eat the milk of carnivores for cultural or psychological reasons, or because meat eating badly flavors milk. But a pig is whatever you make it. Pigs will eat anything, and they can be vegetarian if you prefer. Perhaps we shun pig milk because we prefer to drink from animals that bear one to three babies and have their teats arranged in a single bladder, an udder.

Northern Europeans once considered reindeer milk the best milk of all, and for a time were also partial to elk milk. Neither has remained popular.

Comparing the different types of milk is complex. But in the beginning, the most important issues involving milk were simple:

What milk-producing animal was both easiest to domesticate and available in large numbers?

All evidence indicates that milking animals began in the Middle East, possibly in Iraq or the Assyrian part of Iran. Sumerians in the city of Ur created a frieze on a wall of the temple of al-Ubaid five thousand years ago of a scene of dairy workers milking cows and pouring the liquid into large jars. But as early as was this frieze, known to archeologists as "the dairy of al-Ubaid," it probably did not depict the earliest milking, because cows were probably not available when milking began. Civilization in this area between the Tigris and Euphrates rivers is thought to date back seven thousand years.

67. MILKING OF THE REIN-DEER.

Designed and Engraved by Messrs. Sly and Wilson. The Animals from living Specimens, and the Accessories from De Broke's 'Lapland.'

Milking of the reindeer in Lapland, the northernmost region of Finland.
An engraving from The Art-Union Scrap Book, *London, 1843.*
Artists Sly and Wilson. (HIP/Art Resource, NY)

Man milking cow in detail of a relief on a limestone sarcophagus of Queen Kawit, wife of Pharaoh Mentuhotep II. Egypt, Middle Kingdom, 2061–2010 B.C. From Deir el-Bahri, Werner Forman Archive, Egyptian Museum, Cairo, Egypt.
(HIP/Art Resource, NY)

Archaeological finds suggest that humans have been herding animals for ten thousand years, and they must have been living close to them for at least that long because animal pathogens started mutating into human diseases such as smallpox, measles, and tuberculosis ten thousand years ago. Was it then that milking started?

No one really knows. How was it decided that the milk of pastoral mammals could be substituted for human milk when a mother died or was unable to produce enough milk? It seems a bold step to replace mother's milk with that of an animal.

But perhaps animal milk was first recognized as a commercial product and only later used for feeding human babies. In a hot climate where milk spoiled very quickly, cheese and yogurt, made from soured milk, must have been developed early. In fact, until the age of refrigeration, very little fresh drinking milk was consumed in the Middle East.

Or perhaps the practice of humans drinking other mammals' milk began when lactating animals were used as wet nurses, the human babies placed on teats to suck milk. This practice occurred in ancient times, in the Middle Ages, and even in more modern times in poor parts of Europe. It is not known how frequently it actually occurred, but it is striking how often it comes up in the literature and mythology of Egypt, Greece, and Rome.

In ancient times, when abandoning babies was commonplace, there were many stories of infants being saved by lactating animals. The symbol of Rome is a depiction of its two founders, the twin boys Romulus and Remus, breastfeeding on the wolf that raised them.

Another mystery is what kind of animal was first used for milking. It was almost certainly not a cow. If milking really did start in the Middle East ten thousand years ago—or even nine or eight—it must have been with some other animal, as there were not many cows there then, or anywhere else, for that matter.

The ancestor of cattle, of all bovines, was the aurochs. Aurochsen (the correct plural form) were large, powerful, violent, and aggressive animals. With horns more than two feet long and shoulders that stood higher than the height of a man, they fearlessly attacked the humans who hunted them and inspired awe, as testified to by the frequency with which they appear in cave wall paintings. Females were probably less aggressive than males, but even so, milking a wild aurochs was as practical as trying to milk a wild bison on the plains of North America. In all likelihood, neither ever happened.

In time, the wild aurochs were domesticated, and as this domestic version proliferated, the wild aurochs began to disappear. Once ranging over an area stretching from Asia through Europe, they eventually were confined to the forests of central Europe. The last aurochs died in seventeenth-century Poland.

Modern cattle do not stem from these last Central Europeans, but from a cousin, the Urus, which was very hairy and, according to Caesar, almost as large as an elephant. The Urus roamed over Europe, Asia, and Africa. An early domesticated Urus breed was the Celtic shorthorn, a small but sturdy animal that not only provided the Celts with milk, but was the ancestor of many modern breeds.

Was the first milking animal a goat, as goat enthusiasts always claim? Or was it a gazelle, the wild ancestor of goats? This is possible, but gazelle farming would have been difficult unless they were soon domesticated into goats. Perhaps it was a sheep, a relative of the goat. But sheep's milk, with its high fat and protein content, is rich for drinking, and sheep produce only miserly amounts of milk.

Still, the Sumerians' first dominant domestic animal was the sheep, which they began domesticating six thousand years ago. The value of Sumerian sheep was not appraised by their capacity to give milk, but by the size of their tails. The ample fat found in sheep tails was an important source of cooking oil.

According to Sumerian tablets, a typical sheep flock numbered between 150 and 180. But some flocks were as large as five hundred. The sheep seem to have been well cared for, especially those owned by the clergy, who had special grazing fields and supplemented their sheep's diet with dates and bread. They did, after all, want them to have fat tails.

The Sumerians also raised cows and goats, though in much smaller numbers than the sheep, fueling further speculation as to which animal came first. But whatever the case, the Sumerians seemed to struggle with a declining milk production of their domestic stock. They kept crossbreeding them with wild animals—cows with bison, goats with wild mountain goats, and sheep with wild rams.

The shepherds used whatever milk was produced to make butter, cream, and several types of cheese. Soured milk with honey was used as a cough remedy. But the consumption of dairy products was not widespread, and what there was of it was controlled by the temple.

Camels could also have been the first animal milked by humans. Camels are convenient to milk because they are tall animals—though somewhat grouchy. They have the advantage of being able to find food anywhere. They graze in desert areas seemingly devoid of forage, nourishing themselves on scrubby plants that other animals won't touch and humans often don't even notice. Camels were milked in the Middle East, but it is not known when that practice began. Pliny opined that camels had the sweetest milk, although if judged by lactose content, that would not be true. The llama, closely related to the camel, provides milk in South America today, but these animals were not milked until the Europeans arrived.

The British writer Isabella Beeton, in her bestselling 1861 book, *Mrs. Beeton's Book of Household Management*, arrived at an assessment of milks that remains a fairly good summary today:

> Milk of the human subject is much thinner than cow's milk; ass's milk comes the nearest to human milk of any other; goat's milk is something thicker and richer than cow's milk; ewe's milk has the appearance of cow's milk and affords a large quantity of cream. Mare's milk affords more sugar than that of the ewe; camel's milk is employed only in Africa; buffalo's milk is employed in India.

Once cows became easily available, most milk producers chose to milk them rather than other animals, though that choice has never been without controversy. Mohandas Gandhi, father of the modern cow-worshipping state of India, drank exclusively goat's milk, which he considered most healthful. But cows are easy to work with and they produce a tremendous amount of milk. A goat might produce three quarts in a day; a really good goat, a gallon. A cow naturally produces several gallons a day, and modern farmers using advanced production techniques can hope for eight or more. However, the larger the animal, the more it has to be fed, and a goat produces five times as much milk in proportion to her body weight as a cow, four times as much milk for her weight as a sheep.

Goats have another advantage over cows, especially in the Middle East and North Africa. They don't need rich green pastures in which to graze and can find food in places where a cow would starve. They can even climb trees to eat leaves.

Farmers need animals with whom they can have a peaceful, even affectionate, relationship. The Assyrians chanted prayers and magical invocations in which they asked for their livestock to have a friendly attitude.

There is a simple trick for more easily milking cows, goats, sheep, and other animals that animal rights activists abhor. When a calf, kid, or lamb is born, it is taken away from its mother and fed milk from a bottle by the farmer. Most animal rights activists say that the separated animals moan and cry with grief. Some farmers say they do, too. Others deny it or appear not to care; as Brad Kessler, a small-scale Vermont goat farmer puts it, "Milk is power." Separating a calf from its mother is another of the many enduring controversies around milk.

When animals are left to suckle their mothers, they drink up a considerable portion of their milk—more than they need—and a farmer's profits. They also grow up to be very independent and some-times distrustful of humans. But if a farmer feeds the young animals, they grow up with a real fondness for human beings. Cows are too big to frolic with people the way goats sometimes do, but they nuzzle farmers with their noses and follow them around. They like life to be calm and easy, which is why cow farmers are usually calm,

soft-spoken people. Sheep follow farmers around in a cluster but demonstrate none of the individualistic behavior of cows or goats. They seem to exist more as a flock than as individuals, which is probably why there is no distinct name for them in the singular. Farmers and milking animals can enjoy a very warm relationship, but it never ends well for the animal, because a farmer cannot afford to keep feeding an animal that has stopped producing milk.

2

GOING SOUR IN THE FERTILE CRESCENT

FROM ANCIENT TIMES up to the modern era, there have always been babies who were breastfed and babies who were bottle-fed. And from the very beginning, the relative merits and disadvantages of each have been fiercely debated.

Today, many advocates of breastfeeding insist that there are relatively few women who cannot produce milk. But records dating back to the beginnings of civilizations reveal that even back then there were many formulas and medicines for "failed" mothers. The Ebers papyrus, so called because it was purchased in 1873 by a George Ebers, is a 1550 B.C.E. papyrus scroll on herbal medicine that offers a prescription for struggling mothers: Rub the woman's back with oil and warmed swordfish bones.

If, for some inexplicable reason, the oil and swordfish bones didn't resolve the problem, the papyrus recommends hiring a wet nurse. Wet-nursing was often seen as preferable to "artificial feeding," the term by which feeding infants animal milk was known for centuries, but not always—another enduring controversy.

Wet nurses needed to be watched carefully and regulated with great seriousness, as was made clear in the Babylonian legal code of Hammurabi circa 1754 B.C.E. The code stated that if a baby died while in the care of a wet nurse, she could not try to sneak in another baby in its place. Any nursemaid caught doing so would have her breasts cut off.

Sometimes a family sent a baby to a wet nurse but did not have the money to pay her. The code allowed that in such a case, the family could sell the baby to the wet nurse.

The legal codes of the classical world—Egypt, Greece, and Rome—all had laws making it clear that wet-nursing was a highly respectable profession governed by legal contracts. When the baby Moses was found on the banks of the Nile, the pharaoh's daughter, according to the legend, hired a lactating woman and ordered her to feed the child. The fact that she did this rather than feeding the child animal milk has always been interpreted as evidence of her determination to raise the baby with special care.

In most representations of nursing children, the child is being fed from the left breast, which was thought to contain the best milk because it was closest to the heart; it was only with the introduction of Christianity that this belief began to fade. An aristocratic baby had to be fed exclusively from the left breast. This was possible because highborn babies, especially in Egypt, often had several wet nurses.

Nursemaids were also considered to be a better option than feeding a child animal milk, because in a hot climate, milk can be dangerous. It has to get from the animal to the baby quickly, before deadly bacteria sets in, and in the classical world, this could be a difficult task. Nevertheless, we know that at least some babies were bottle-fed early on because the ancients left bottles behind. Egyptian terracotta nursing bottles dating from 1500 B.C.E. have been found, as well as Egyptian nursing vessels of various designs going back to 4000 B.C.E.

In ancient times, as in modern times, there were women who did not want to breastfeed. The fact that they were usually highborn women suggests that only those with some financial means could afford to hire a wet nurse or to rush fresh animal milk to their babies. It may also have been that some social status was needed to refuse a task that many regarded as a woman's obligation.

In the Egypt of the pharaohs, wet nurses lived in a harem and were well treated and highly regarded. Their names were placed on

Stone receptacles for milk from Vanous, Cyprus, ca. 2200–2100 B.C. Cyprus Museum. (SEF/Art Resource, NY)

important party and funeral lists. In Greece by 950 B.C.E., it was fashionable for upper-class women to hire lower-class nursemaids. Wet-nursing was also often the work of slaves, and thus slave-owning women did not breastfeed.

The option of bottle-feeding, in contrast, may have been resorted to by women who had no other choice—overworked and underfed poor women who could not produce enough milk. Bottles were also probably used for the babies of poor women who died in child-birth, and for abandoned children shunted off to orphanages. To substitute animal milk for human milk may have been a last, desperate measure.

Elaborate cups formed in the shape of a woman nursing a baby or holding out a breast from which the milk poured seem to have been the nursing cups of choice for more affluent babies. Poor babies were fed from animal horns.

IT SEEMS LIKELY that milk was not originally produced for feeding babies or for drinking. Instead, as an extremely unstable product, it was probably cured, hardened, soured, or fermented into a variety of highly nutritious and stable foods.

Many centuries before Louis Pasteur, the ancient Assyrians knew, probably from their own experience, that the only way to keep fresh milk from becoming poisonous was to boil it. The resulting scum on the pot mixed with breadcrumbs was a children's treat, which they lapped directly from the pot. It was believed back then, and many twenty-first-century people would agree, that boiled milk lacks flavor and that only the scum and the skin left on top are good to eat.

Yogurt is made by adding a live culture to milk, another trick apparently learned in very ancient times. The milk was then boiled, and the pot was wrapped in cloth and left to cool very slowly, first indoors and later outside in the night air. But when the yogurt was outside, it had to be guarded carefully, as many animals, especially cats, are great yogurt lovers. Thick and sour yogurt gives off an inviting fragrance. In some places, yogurt is still made this way today.

Butter, which is a well-preserved milk fat, was made by shaking cream in a goatskin. In the Middle East, butter was never spread on bread, though there was a tradition of dipping bread in butter. Instead, butter was used for making holiday dishes throughout the year. To this end, the butter was salted, because only salted butter lasts. Unsalted butter was a luxury usually only available after refrigeration was invented. And even then, until recently, many stores, including those in America, were unwilling to stock unsalted butter because it would not sell fast enough to prevent spoilage.

The frieze of the ancient dairy of Ur shows men rocking a clay urn to make butter. And the Hittites used butter because they could produce it for half the cost of olive oil, which was probably considered to be a higher-quality product. Adding to the low reputation of butter, it was probably often consumed in an already slightly rancid condition.

When the butter was churned, whether in an urn or a goatskin, the remaining liquid was what we now call buttermilk. It was probably a rich buttermilk, because separation techniques were crude; the buttermilk often contained small lumps of butter. When buttermilk was abundant, it was fed to farm animals. It was also a popular drink among peasants. Buttermilk made an appearance in the cities only when country people moved in.

The high regard in which the ancient Assyrians held their livestock and the milks they produced is apparent in customs that continued until a generation or two ago. A supplicant gave the church a fattened sheep, and gave to charity the milk of a first milking, and the butter and cheese produced by that milking. A "first milking" refers to the first milk produced by a cow after she gives birth, usually in the spring.

THE PART OF the world where dairying began did not favor cows. Cows are not a hot-weather animal. They enjoy cooler climates that produce rich green grasslands. This in itself is testimony to how much the early peoples must have preferred the mild milk of cows to the milk of other animals, because in early times, and still today, dairy cows were raised in climates where they did not belong. Birthing takes place in the spring, so animals lactate in the spring and summer, the worst possible time in the hot climate of the Middle East. A freshly drawn bucket of milk starts growing dangerous bacteria within minutes—especially when buckets are not particularly clean, as was probably often the case in early times.

Everywhere in the Middle East and around the Mediterranean, people tackled this problem by turning milk into milk products as soon as possible. The most common product they made was yogurt, though until modern times, no one used this word. The Persians, among the earliest yogurt enthusiasts, called it *mast*, and the food is so central to Persian culture that it frequently comes up in popular idioms. There is an expression for mind your own business, *Boro mastetobezan*, that translates into "Go beat your own yogurt." Or, if something is really scary, people may say *mast-roo sefid shod*, meaning "The yogurt turned white." Yogurt is sometimes a drink, sometimes eaten with a spoon, and sometimes poured over or cooked into meat dishes as a sauce or stew base.

Yogurt drinks were, and are, popular worldwide, and they have different names in different countries: *doogh* in Iran, where salt and mint are added, *lassi* in India, where it may be sugared and salted, and *laban* in the Arab world.

In old Persian, *doogh* meant milk. But as a drink, it was probably always soured, and eventually the word came to mean "yogurt liquefied with water." In modern times, *doogh* is made with carbonated water. In the twentieth century, the Persians started bottling *doogh* with the spring water from Cheshmeh-Ab-e Ali, Ali's Spring. Ali was the Prophet Muhammad's seventh-century son-in-law, and he supposedly commanded this water to spring forth from the barren foothills of the mountains outside Tehran.

In Persia, *doogh* was also sun-dried. The resulting product, called *kahshk*, was then ground into a powder, moistened with water, and rolled into balls. *Kahshk* is still used as a kind of seasoning. The balls are dissolved in soups or stews to give the tangy, sour flavor of yogurt without the texture.

3

CHEESY CIVILIZATION

THERE ARE A great many apocryphal stories of how milk first came into contact with the butchered innards of an animal, specifically the stomach lining, where the milk quickly curdles. The most ubiquitous stories are of nomads traveling with milk stored in bladders made of animal stomach and then, arriving at their destination, finding that the milk had turned solid. The agent from the animal's stomach that curdles the milk is known as rennet. The proteins in milk cannot fuse because they are negatively charged, the way the two negative ends of a magnet repel each other. An enzyme in rennet takes away the negative charge and the proteins start to fuse into curds.

To make cheese, the curds were placed in wooden molds with holes and pressed for days until 85 percent of the liquid—a cloudy, highly nutritious water called whey—was squeezed out of the curds. The rennet leaves with the whey. Whey was, and still is, often fed to farm animals. Many human foods were, and are, made from it as well. The ancient Persians used whey to whip and cook into a food they called *qaraqorut*.

The solid cheese that remains after the whey is removed can then be preserved through brining to produce a cheese like the Greek feta, which is one of the oldest cheeses in the world. Later, many other, more sophisticated variations of cheese were developed in Europe, where damp, cool cellars were available for aging.

The fact that dairy farmers eventually settled on milking cows, goats, and sheep rather than other animals shows the importance of cheesemaking to the early dairy farms. It was generally agreed that those animals produced the best cheesemaking milks, though the argument about which of the three was best has never been settled.

It is not clear when cheesemaking began. Butter and yogurt were probably produced earlier because they are easier to make. Ancient people wrote of eating curds, but it is not always certain what they meant by that. There are numerous references in the Bible to something that might have been either butter or curds.

Curds is more likely, because Mediterranean people had little need for butter. They already had olive oil, which is less prone to spoilage, heats to much higher temperatures without burning, and was and is regarded as more healthful. Even now in North Africa, most of Greece, Mediterranean France, and Spain, and most—but certainly not all—of Italy, olive oil dominates and butter is rarely used. An omelet may be made with butter in Greece today, but until recently, even that was made with olive oil.

The Thracians, the people who lived to the north of Greece, ancestors of the Bulgarians and others, ate butter. Even farther north, the Germanic people were also avid butter eaters. Butter is easier to work with in cooler climates, and the Germans were said to have perfected salted butter.

Similarly to the Japanese Buddhists who called Westerners "butter stinkers," Greeks contemptuously referred to the Thracians as "butter eaters." And the word "butter" itself was not spoken kindly. The Greeks called it *boutyros*, cow curds. They were sheep and goat people, and they regarded those who kept cows and made butter as an alien lot. The Romans thought butter was a useful ointment for burns but didn't regard it as a suitable food. As Pliny the Elder bluntly put it, butter is "the choicest food among barbarian tribes."

The Mesopotamians and Hittites made cow, goat, and sheep's milk cheese, as did the ancient Egyptians. The Greeks were also cheesemakers, and it is deduced from four-thousand-year-old cups with images of cows and goats on them that the ancient Cretans were milk drinkers and cheesemakers, too.

In Greek mythology, Aristaio, son of Apollo, invented cheese, which suggests that the Greeks thought cheesemaking important. The ancient Greeks also heated and slowly cooled cream to create a kind of thick, clotted cream that they mixed with honey and served with game birds. And, like the Persians, they made yogurt, to be eaten plain or with honey, and a kind of milk dessert pudding called *khórion*.

The Greeks used soured milk to produce other yogurtlike products as well. One, called *oxygola*, was said to be fairly hard in texture. Its whey was drawn out and then salted and sealed in a jar. Another, *melca*, was made from liquid milk soured in jars of boiling vinegar and cured overnight in a warm place. The celebrated first-century A.D. cook Marcus Gavius Apicius gave this *melca* recipe; basically, it is another milk-and-honey concoction:

> Mix melca either with honey and brine or with salt oil and chopped coriander.

The stories of rennet's being discovered through the accidental commingling of meat innards and milk is further shaken by the fact that the ancient Greeks made rennet from fig sap. The sap was used to produce aged cheeses and hard cheeses, which were sold in marketplaces in Athens separate from the ones that sold "green" cheeses (soft, fresh cheeses).

The not infrequent references to cheesemaking in Homer present it as commonplace activity on farms. Homer, whose writings were compilations of already ancient oral histories, twice mentions a dish made of barley meal, honey, Pramnian wine (a strong, dark, high-quality wine), and grated goat's-milk cheese. Cheese was also central to the Spartan diet. A rite of passage for a boy was to steal a cheese from a household without getting caught.

Mixtures of grain and dairy, usually resulting in some kind of porridge, were commonplace in both Greece and Rome. Greeks made milk-and-grain porridge on the street in pots of eight or more gallons.

Tracta, a forerunner of pasta, was flour and water fashioned into different shapes—balls, strings, sheets. They were often cooked in milk. Apicius offered this recipe in which *tracta* thickened milk to make a sauce for lamb:

> Put a *sextarius* [about one pint] of milk and a little water in a new cooking pot and let it boil over a slow heat. Dry three balls of *tracta*, break them up, and drop the pieces into the milk. To stop it from burning, stir it with water. When it is cooked, pour it over lamb.

Apicius also suggested a chicken dish made with *tracta*-thickened milk:

> Lift the chicken when it is cooked from the stock and put it into a new cooking pot with milk and a little sauce, honey, and a minimum of water. Place by a slow fire to warm, break up the *tracta* and add gradually. Stir constantly to prevent burning. Put the chicken into this.

This type of flour-thickened milk sauce would eventually become a mainstay of classical French cooking.

THE ANCIENT ROMANS also made and cooked with cheese. Marcus Porcius Cato, commonly referred to as Cato the Elder, who lived from 234 to 149 B.C.E., was a conservative Roman politician from an agricultural background who opposed what he saw as a growing tendency toward excessive luxury. His treatise on agriculture, *De Agricultura*, is the oldest surviving complete book of Latin prose. In it he offers a number of recipes. Here is his *mustacei*, a simple recipe for a soft fresh cheese made with lard and unfermented must, the juice of wine grapes:

Prepare *mustacei* thus: Moisten a *modius* [two gallons, a peck, or a half bushel] of fine flour with must. Add anise, cumin, two pounds of fat, 1 pound of cheese, and a grated bay twig. When you have shaped them, place bay leaves beneath and cook.

Cato's most famous recipe is for placenta, a kind of cheesecake used in religious ritual.

Make placenta this way: two pounds bread-wheat flour to make the base; four pounds flour and two pounds emmer groats [husked grain flour to make layers; in Latin he uses the word *tracta*]. Turn the emmer into water; when it is really soft turn into a mixing bowl and drain well; then knead it with your hands and when it is well worked add the four pounds flour gradually and make into sheets. Arrange them in a basket to dry out.

When they are dry, rearrange them neatly. In making each sheet when you have kneaded them, press them with a cloth soaked in oil, wipe them round and dampen them.

When they are made, heat up your cooking fire and your crock. Then moisten the two pounds flour and knead it: from this you make a thin base.

Put in water 14 pounds sheep's milk, not sour, quite fresh. Let it steep, changing the water three times. [Cheese was stored in brine and had to be soaked to be desalinated.] Take it out and squeeze it gradually dry with the hands. When all the cheese is properly dried out, in a clean mixing bowl knead it with the hands, breaking it down as much as possible. Then take a clean flour sieve and press the cheese through the sieve into the bowl. Then take four and a half pounds good honey and mix it well with the cheese.

Then put the base on a clean table that has a foot of space. With oiled bay leaves under it make the placenta.

First put a single sheet over the whole base, then one by one, with the mixture, add them spreading in such a way that you eventually use up all the cheese and honey and on the top put one more sheet by itself. Then draw up the edges of the base having previously stoked up the fire; then place the placenta to cook, cover it with the heated crock and put hot coals around and above it.

Be sure to cook it well and slowly. Open to check on it two or three times. This makes a one-gallon placenta.

In Rome, cheese was eaten by both the rich and the poor. A considerable variety of hard, soft, and smoked cheeses were produced in the city, and others were imported from around the empire.

The Romans were fond of smoking and made smoked meat, sausages, and other foods, including cheese. Smoked goat's-milk cheese from Velabrum, the valley by the Forum that runs up to Capitoline Hill, one of the seven hills of Rome, was especially popular. The cheese was sometimes eaten warmed or grilled.

We do not know the exact nature of all of these cheeses, but we do know from the few recipes left behind that the ancient Romans made cheese much the way we still do. Columella, in his extensive first-century A.D. writings on agriculture, gave instructions for cheesemaking. First, he added a pennyweight of rennet to five quarts of milk. The milk was then gently warmed until it curdled, strained through a wicker basket, and pressed in a mold. Next, it was either salted or soaked in brine. This has remained a way to make cheese to this day.

Like the Greeks, the Romans made rennet from figs. They also made it from artichokes, and from the innards of sheep, goats, donkeys, and hares.

The Romans liked fresh cheeses as well as aged and smoked ones. Curds only a day old were salted and seasoned with local herbs such as thyme or crushed pine nuts. Sometimes the seasoning was added to the milk bucket before the cow was milked.

Cheeses were often given as gifts, and they were a standard breakfast food, along with olives, eggs, bread, honey, and sometimes leftovers from the night before. Cheese was also eaten for lunch and dinner, sometimes as an appetizer and often as a dessert, which was said to be bad for digestion.

Because fresh milk was available only on farms, it was consumed mostly by the farmers' children and by peasants who lived nearby, often with salted or sweetened bread. This led to fresh milk's being widely regarded as a food of low status. Drinking milk was something that only crude, uneducated rural people did and was rare among adults of all social classes.

The Romans, who often commented on the inferiority of other cultures, took excessive milk drinking as evidence of barbarism. Because milk spoiled quickly in the climate of southern Europe and kept far better in northern Europe, northerners used far more milk. This led southern classical cultures, which were contemptuous of northerners in any event, to take the greater consumption of dairy as evidence of their barbarian nature. During a visit to conquered Britain, Julius Caesar was appalled by how much milk and meat the northerners consumed. Strabo disparaged the Celts for excessive milk drinking and excessive eating in general. Tacitus, when illustrating the crude and tasteless diet of the Germans, singled out their fondness for "curdled milk."

Anthimus, a sixth-century Greek living in exile in Ostragoth, cautioned that milk was more healthful when not drunk fresh:

> If it is for people with dysentery, give goat's milk prepared by heating round stones in the fire, and then putting these stones into the milk. When the milk boils, take out the stones. Add white well-leavened bread that has been finely sliced and cut up into pieces and cook slowly on the fire. Use an earthenware pot and not a bronze pan. When the milk has boiled and after the bread has become saturated, let the patients eat the pieces with a spoon. Milk is more beneficial served like this, because it is nourishing. If milk is drunk on its own, it passes in contrast straight through and scarcely remains in the body.

Anthimus believed that while fresh cheese was harmless, especially if dipped in honey, cured cheese caused kidney stones and was very unhealthful. He wrote, "Whoever eats baked or boiled cheese has no need of another poison." He was neither the first nor the last to warn of cheese. Hippocrates, the fifth-century B.C.E. Greek who is considered the father of medicine, also warned about cheese:

> Cheese does not harm all people alike, and there are some people who can eat as much of it as they like without the slightest adverse effects. Indeed it is wonderfully nourishing food for the people with whom it agrees. But others suffer dreadfully . . .

So began a twenty-five-hundred-year-long debate about the health merits of cheese. Many have given opinions. According to Aulus Cornelius Celsus, the first-century author of *De Medicina*, aged cheeses contained "bad juices that were harmful to the stomach." Bartolomeo Sacchi, commonly known as Platina, a writer on food at the height of the fifteenth-century Florentine Renaissance, stated that fresh cheese was very healthful except for people who were phlegmatic, but "aged cheese" was "difficult to digest, of little nutriment, not good for stomach or belly, and produces bile, gout, pleurisy, sand grains, and stones." However, he did add that a small amount of cheese after dinner benefited digestion.

A suspicion of cheese has never gone away. In the nineteenth century, Alexandre Dumas wrote, "It is the coarsest part of milk and the most compact, so obviously it produces a substantial food stuff, but one that is difficult to digest if eaten to excess." Modern nutritionists also warn against eating cheese in excess because of its high fat and cholesterol content.

IN THE ENDLESS arguments over which animal had the healthiest milk, Anthimus favored goats, while Marcus Terentius Varro, a first-century B.C.E. Roman writer, favored sheep. Galen, the prominent Greek physician in third-century A.D. Rome, thought goat's milk most nourishing but cow's milk more medicinal and more soothing for the sick. He also praised sheep's milk as the sweetest. Wealthy Roman women used donkey's milk as a cosmetic because it was supposed to both smooth out wrinkles and make skin whiter.

Most everyone put cow's milk—and, by extension, cow cheeses— a distant third. But in the latter part of the Roman Empire, cow's-milk cheeses became increasingly popular. Apicius specifies cow's-milk cheese in some of his recipes.

It was commonly believed that milk was bad for the teeth. Today we believe that milk is good for the teeth because of its high calcium content, but how much calcium is healthful is another long-running debate. The Roman idea that milk is harmful to teeth persisted for centuries. The seventeenth-century English physician Tobias Venner advised his readers to gargle with wine or strong beer after drinking milk.

Anthimus stated that the only really salutary way to drink milk was while it was still warm from the udder. This idea persisted for many centuries, until the age of refrigeration. Many people preferred to go to a farm to drink milk, or sometimes had the animal brought to them. In numerous cities from London to Havana until the twentieth century, milk deliveries were sometimes made by leading a cow from door to door.

Many believed that milk needed to be consumed on an empty stomach because food in the stomach caused curdling, which could be dangerous. Galen's remedy for this was to always mix fresh milk with honey, though he admitted that that didn't work for everyone. He also advised using other additives, such as salt and mint.

Galen's extensive writings on milk apparently influenced many who came after him, including Platina, the Renaissance food writer, who in turn influenced other people over the next few centuries. For example, Platina's advice not to drink milk at the end of the meal was reiterated a century later by the Spanish medical writer Francisco Núñez de Oria.

Platina's opinions about drinking milk mirror Galen's:

> [Milk] is better in spring than summer, and better in summer than autumn or winter. It ought to be drunk on an empty stomach, warm as it comes from the udder, and one ought to abstain from other food while it settles in the stomach. It is least harmful drunk as curd at the first course in spring and summer, for taken after the meal, as we are generally accustomed to do, it either spoils immediately or draws other undigested food with it to the bottom. One should also be quiet after taking it so that it will not sour in the stomach from shaking . . . One must, however, avoid too much use of milk, for it makes the keenness of the eyes duller and generates stones in kidneys and bladder.

In Roman writing there were frequent warnings about excess, because they loved excess, especially in the first and second centuries. They had a popular cheese spread called *moretum*, which means salad. The name suggests that the spread was more vegetable than cheese, except that the vegetable in question was garlic. A late-first-century poem titled "Moretum"—historians debate its author—describes the preparing of the dish. It does not specify the amount of cheese, but states that the four heads of garlic should be separated, peeled, and crushed before being combined with the cheese, undoubtedly resulting in an extremely powerful garlic cheese spread. The poem states:

> *The vapor keen doth oft assail the man's*
> *Uncovered nostrils, and with face and nose*
> *Retracted doth he curse his early meal;*
> *With back of hand his weeping eyes he oft*
> *Doth wipe, and raging, heaps reviling on*

The undeserving smoke. The work advanced:
No longer full of jottings as before,
But steadily the pestle circles smooth
Described. Some drops of olive oil he now
Instills, and pours upon its strength besides
A little of his scanty vinegar,
    And mixes once again his handiwork,
And mixed withdraws it: then with fingers twain
Round all the mortar doth he go at last
And into one coherent ball doth bring
The different portions, that it may the name
And likeness of a finished salad fit.

This concoction, no doubt, was startling to the digestive system.

4

BUTTERY BARBARIANS

THOUGH CHRISTIANITY BEGAN in the Middle East and only later spread to the dairy country of northern Europe, it has always been a milk-worshipping faith. In the early communions, the sip that represented the blood of Christ was often taken from a goblet of milk.

According to many authorities, using milk rather than wine to represent the blood of Christ was logical because milk was believed to be a form of white blood. This idea long predated Christianity, but the early Christians seized upon it. In a lengthy 198 A.D. treatise entitled *Paedagogus*, Clement of Alexandria, an early Christian theologian, presented several arguments for the use of milk in Christian ritual. He opined that a mother's milk was blood that had been sweetened and purified, and that semen was blood that had been whipped into a foam.

Clement also pointed out that in 1 Corinthians 3:2, Paul compares his teachings to milk and meat—nourishment. Clement wrote, "The blood of the Word has been also exhibited as milk."

Clement proposes an unappetizing formula: mixing milk with wine as a means to touch immortality. The wine would curdle the milk and the whey could be drained off, just as the drawing off of lust and other impure thoughts could lead a man or woman to eternal life.

It is not difficult to imagine why wine mixed with milk did not become an established ritual. Nonetheless, the use of milk instead of wine in religious ceremony stubbornly endured despite Pope Julius I's condemnation of it in 340 A.D. Similarly, the practice of feeding a baby milk with honey for its first communion continued well into the Middle Ages. Each of the foods used in the rituals had significance: bread for the body of Christ, wine for the blood of Christ, and milk and honey for the Promised Land.

Even in the Middle East, the Christian tie to milk never completely disappeared. There was said to be a cave in Bethlehem where the Virgin Mary had nursed Jesus and spilled a drop of milk. Women who were barren or could not produce milk went there for help.

Perhaps it was not a coincidence that just as Pope Julius was banning the use of milk in ritual, Christianity was spreading to the

Madonna of the Milk *by Ambrogio Lorenzetti (ca. 1311–1348) depicts the Virgin Mary breastfeeding Christ. Oratorio di S. Bernardino, Siena, Italy.* (HIP/Art Resource, NY)

dairy-loving barbarians. Saint Brigid, the fifth-century patroness of milk- and butter-loving Ireland, said to have been nourished as a baby by a white-and-red cow, was the revered saint of dairy farmers, cattle, and milkmaids in the newly Christian north.

In the early Middle Ages, goats and sheep were preferred to cows except in a few regions such as the Alps, and liquid milk, fragile and unstable as it is, even in a cool climate, was seldom commercialized. Cheese and butter were the primary dairy products, and cream was almost always churned into butter. The by-product, buttermilk, was very popular.

Ancient and medieval butter was often buried in order to ferment it slightly. In Ireland, butter was buried in peat bogs. Before industrial dairies, butter was not entirely fat, no matter how well churned the cream was, and even today, butter is usually between 75 percent and 85 percent fat. French butter makes better pastry than American butter because it contains more fat and less water.

STRABO'S OBSERVATION ABOUT the Celts, that they were a dairy-eating people, has remained true to this day. But the Celts were not among the butter-eating barbarians who took over Rome; that victory belonged to the Franks, Vandals, and Goths, restless tribesmen who lived on meat, milk, and cheese and roamed about in constant search of new pastureland.

The Celts came from the upper Danube and, by the fifth century B.C.E., controlled much of northern Europe. But in time they were driven out of their ancestral homeland to Europe's Atlantic edges, which proved good places for dairying. They settled down in what is now Scotland, Ireland, Wales, the Isle of Man, the Breton coast of France, and, with less of a cultural connection, northwest Spain. They also became known for their butter. The Irish boiled milk with a seaweed known as *cairigin* or Irish moss and sweetened it with honey.

The English, whose model for imperialism was the Romans, sneered Roman-like at what they thought was the barbaric overuse of butter by the Irish. Fynes Moryson, secretary to the viceroy, who spent much time in Ireland in the reign of Elizabeth I, reported that the Irish "swallow whole lumps of filthy butter."

In his 1672 travel book, the geographer and mapmaker Albert Jouvin de Rochefort wrote, more generously, "Butter is in Brittany, like in nowhere else in France, a centerpiece of a rich and ancient cultural tradition." It was also often observed that the Bretons, or people who live in Brittany, ate far more butter and far less cheese than the French, who were themselves butter eaters, at least in the north.

The Bretons were butter gourmets and had the unwelcome habit of pinching a sample before buying. Celts, and specifically Bretons, were also salt makers. They always salted their butter, and they had very strong opinions about what was undersalted or oversalted. They used butter molds made of carved wood and brought specially molded decorative butters to weddings and funerals.

One of the great expressions of the Celts' love for butter is the many butter cakes that are made throughout the Celtic world. These are simple cakes designed to taste as buttery as possible. Brittany is especially known for two versions, the *kouign amann*, which in the Breton dialect of the Celtic language means "butter cake," and the *gâteau breton*, which in French means "Brittany cake." The *gâteau breton* appears to be the more traditional of the two, as it uses the buckwheat flour that is typical of Brittany, many eggs, and more butter than does the *kouign amann*, and it is baked in a cast-iron skillet, like an old Celtic cake. But there is no record of it before the Paris Exposition of 1863, which is about the time the baker Yves-René Scordia invented the *kouign amann*.

Butter cakes, along with shortbreads, had been around for centuries before these versions from Brittany, however. An example is the Irish scone. Scones are traditionally cooked on a griddle, which is typical of old Celtic baking, and are probably of Scottish origin; the word "scone" comes from Gaelic.

Here is a recipe, similar to the *gâteau breton* but formed into smaller cakes and made with white flour, from Florence Irwin of County Down on the Irish Sea in what is now Northern Ireland. At the beginning of the twentieth century, she traveled around County Down teaching "domestic science" and recording and trying the old traditional recipes of her students. Note the quantity of butter, probably salted, and the touch of buttermilk used in this Celtic standard. Also note that despite the already buttery taste of these small cakes, Irwin suggests splitting them open on the day they are baked and spreading them with yet more butter. They are to be eaten while they are still warm.

> I lb flour, 4 to 6 ozs butter, 2 ozs castor sugar, good pinch salt, small teaspoonful bicarbonate of soda, small teaspoonful cream of tartar, buttermilk.
>
> Sieve the dry ingredients. Rub in the butter very lightly, handling as little as possible. Using a knife mix to a dough. Turn on to floured board. Very lightly knead. Roll out half to three quarters of an inch thick. Cut in rounds. Brush with buttermilk. Bake in a hot oven till risen and brown.

The original Celtic butter cake would not have had leavening and would have been baked on a griddle, not in an oven. The *gâteau breton* is baked in an iron skillet because that is reminiscent of the iron griddle that came before it. In much of Europe, ovens were not features of homes. Often food had to be taken to a baker to be baked, which is the meaning of the French adjective *boulangère* for certain stews and meat pies that were taken to the baker. But homes had griddles, known in Scotland as "girdles." Originally, a griddle was a flat stone, called a bakestone in England, heated by a fire. Later, the griddles were made of metal. They are still used in some Celtic places for small cakes such as drop scones—the original of the Irish scone above—and Welsh cakes, which in Welsh Celtic are still called *pice ar y maen*, cakes on the stone. Griddle cooking is usually done with buckwheat, barley, or oats, not white flour, and the Scottish also have oatcakes, which are usually made with milk, not butter, and then served buttered.

Welsh cakes were made of barley and milk beaten together into a thick batter that was put in an earthen jug and then poured onto a hot griddle in discs the size of a small saucer. They were only about a third of an inch thick and were still soft when done. They were served with butter.

What does this sound like? Yes, they are pancakes. Pancakes are griddle cakes made with milk or sometimes buttermilk. The first-century Roman cook Apicius made very thin pancakes and served them with honey. And the Breton pancakes may have had a Roman origin, as they, too, are thin and are called *crêpes*, which comes from the Latin word for "curl." All European pancakes, including the Welsh and the Breton versions, are made from a flour-and-milk batter poured on a griddle, but crêpes were originally made with buckwheat.

It is not certain when the Europeans started making pancakes, but they certainly did so before the fifteenth century. In 1615, when pancakes were becoming extremely popular in England, Gervase Markham, a poet and contemporary of Shakespeare, published a cookbook and home guide that became a huge bestseller, *The English Huswife*. In it he stated erroneously that pancakes were more delicate when made with water instead of milk. "There be some who mix pancakes with new milk or cream, but that makes them tough, cloying, crisp." The book sold well, but the English people kept making their pancakes with milk.

The idea of making pancakes with water may have been born out of poverty. William Ellis, an eighteenth-century farmer, was a popular author who wrote about farming, home management, and cooking. In his 1750 book, *Country Housewife Family Companion*, he wrote:

HOW WATER PANCAKES ARE MADE BY POOR PEOPLE

This pancake is made by many poor, day-laboring men's wives, who when they cannot afford to make better, make this: by stirring wheat flower [sic] with water instead of milk for if they can get milk, they generally think it put to a better use when they make milk porridge of it for their family. The flower and water being stirred into a batter consistence, with a sprinkling of salt and flowered ginger, they fry the pancakes in lard, or other fat, and without any sugar they and their families make a good meal of them.

Ellis follows this recipe with one titled "How Pancakes are made for rich people." This second recipe uses cream rather than milk, a lot of butter, and a heavy sprinkling of sugar. This suggests that fresh milk was a luxury item not available to urban poor people. In the eighteenth century, the upper-class English wanted their food to be as rich as possible and in this sense, water pancakes were lighter.

The leading articulator of this rich diet, the last preindustrial British cuisine, was Hannah Glasse, whose books were so popular that Dr. Johnson claimed there was no such person and Hannah Glasse must be the nom de plume of a man. But Hannah did exist, she was a woman, and she cooked with endless quantities of cream and butter. The Romans would have been appalled by all the dairy but would have loved the sense of excess. In her 1747 book *The Art of Cookery Made Plain and Easy*, she offered one recipe for pancakes with milk and five with cream. Here is one. Like almost all Glasse recipes, it is not original, and others were also making pancakes with cream.

FINE PANCAKES

Take half a pint of cream, half a pint of sack [sherry], the yolks of eighteen eggs beat fine, and a little salt, half a pound of fine sugar, a

little beaten cinnamon, Mace, and nutmeg, then put in as much flour
as will run thin over the pan, and fry them in fresh butter.

In Scotland, the grain of choice was oats. The following recipe
comes from the Scotswoman Elizabeth Cleland in 1755, but this way
of making pancakes began centuries earlier (few Scottish recipes were
recorded before the eighteenth century). The recipe's lemon peel,
oranges, nutmeg, and sugar were probably eighteenth-century additions. Also, the original pancakes would have been cooked on a
griddle, not in a pan, and would not have required butter.

Cleland included seven different pancake recipes in her book.
Some used Scottish units of measurement, and it may have been Scottish nationalism that led Cleland to ignore the 1707 Act of Union
that ordered all Scottish measures to be replaced by English ones. She
did switch some measures, but not those related to volume. A *chopin*
is a quart, a *mutchkin* is a pint, and a *gill* is a quarter of a pint:

OATMEAL PANCAKES

Boil a chopin of milk and blend into it a mutchkin of flour of oatmeal,
thus: keep a little milk, and mix the meal by degrees in it. Then stir in
the boiling milk; when it is pretty thick, put it to cool, then beat up six
eggs with sugar, nutmeg, and the grate of a lemon, and a little salt. Stir
all together and then fry in butter putting in a spoonful of the batter
at a time. Serve them up hot with beat butter, orange and sugar.

IN THE MIDDLE Ages in Ireland, as in most Celtic countries, dairy
foods were central to the diet. But they were not always available to
the poor. Only affluent people owned livestock, and butter, buried
in peat bogs to age like a rare treasure, was particularly valued—and
expensive.

Medieval Irish literature praised milk. In *Tochmarc Ailbe*, "The
Wooing of Ailbe," an Irish tale dating back to the eleventh, tenth,
or ninth century—the date is not certain—milk was described as

"Good when fresh, good when old, good when thick, good when thin." The milk of the poor was probably diluted with water so that it would stretch farther. Though milk was probably rarely drunk; rather, it would usually have been thickened with rennet to make cheese, or boiled with herbs and served as a heavy grain porridge.

The Celts made both soft cheeses and harder ones. The harder ones were better for travel and storage. And apparently, sometimes the cheese was very hard. In the twelfth-century Irish story *Aided Meidb*, "The Violent Death of Medb," a man named Furbaide plotted the assassination of Queen Medb thus: The queen bathed in a well on an island every day. Furbaide planted a stick in the ground near the well at just the right height, tied a rope to it of just the right length, and stretched the rope to a spot where he was hiding on the mainland. Then he used the rope of just the right length and the stick of just the right height to practice with a slingshot every day until he had mastered incredible accuracy. And then one day there the queen was, bathing in the well with her head sticking out, her forehead perfectly exposed. With no time to find the perfect stone, Furbaide grabbed the first thing he could find, a piece of hard cheese, and fired. He hit his target perfectly and killed the queen.

To HEAR THE Romans tell it, the barbarians to their north were swilling milk by the mugful. Though Caesar had been appalled by how much milk and meat the northerners ate and drank, in actuality, they were consuming milk conservatively. It was precious, not abundantly available, and they usually used it to make cheese or a broth with herbs. Such broths were also made with ale or only water, so a milk broth was probably a special treat.

Milk was so important to the barbarians that a dry cow was considered to be a family crisis; most families only had one or two cows. A cow was an expensive animal to maintain.

Scots, particularly in the Highlands, were more interested in dairy than in meat. Originally, they had kept sheep, primarily for their wool, but also for their milk. Sheep, however, are not great milk producers. Later, when the Scots started owning cows, they used them for hauling and farm work as well as milking. Only cows that could no longer work or produce milk were slaughtered, meaning that the Scots probably ate a low grade of meat. Quality meat would have been only for the wealthy. Butter was sold commercially, and there was "rent-butter"—tenant farmers paying off landlords in butter. Twice a year, the Scots mixed butter with tar and swabbed it on their sheep to protect them from developing sores.

For centuries, small sheep's-milk cheeses wrapped in seaweed or ashes were the norm in Scotland. The cow's-milk cheeses for which Scotland later became known were only introduced in the eighteenth and nineteenth centuries.

In the Shetland Islands of Scotland there were a variety of dairy dishes with Nordic names, suggesting that the fishermen learned them from the Scandinavians, who had a wealth of dairy dishes. *Blaund* was whey that was strained from buttermilk and slightly fermented so that it had effervescence. *Strabba*, known in Norway as *stoppen*, was whipped curdled milk eaten with fruit. *Kloks* was clotted milk with cinnamon. Cooking it turned it yellowish, and it is said to have resembled modern condensed milk.

There are many traditional dairy foods in the Shetlands. *Beest* is a drink of colostrum with a little water added. It could be curdled by cooking to make a kind of cheese or mixed with sugar, salt, and caraway seeds for a pudding. They so loved buttermilk that in the winter when there was no butter making, they made fake buttermilk by souring potato water.

Milk was scarce, and ways of stretching the milk supply were always sought. As with the Celts, one way to do it was to dilute it with water. In time, this was regarded as fraudulent, but originally it was an openly acknowledged attempt to stretch the milk supply.

The Scottish families also used to stretch their milk by frothing it into what was known as *froh-milk* or, in Gaelic, *omhan*. They used a frothing stick that had a cross at one end with cow tail hairs attached.

The stick would be rubbed between the hands until the milk frothed. A glass of milk could be almost doubled in this way. Sometimes whey was frothed, and in hard times, poor Highlanders lived on almost nothing but this high-protein drink.

Milk shortages continued in Scotland and many other countries until the eighteenth century, when dairying became a commercial rather than a family activity. Only then did milk and other dairy products become affordable commodities.

Both buttermilk and whey were popular drinks throughout Scotland. In Edinburgh, people were passionate about buttermilk. It may seem inevitable that a society of butter eaters would be buttermilk drinkers, but that does not necessarily follow. Certainly not all cheesemakers are whey drinkers, even though the cheesemaking process leaves behind an enormous quantity of whey. Many cheesemakers gave their leftover whey to farm animals, especially pigs, as is still the custom today—and the reason why pigs are often raised on dairy farms.

In Anglo-Saxon society, a female slave was given whey to drink in the summer, probably because this was cheesemaking time and whey was abundantly available. Shepherds were also entitled to a share of whey or buttermilk, and whey was a common payment for labor even after the 1066 conquest by the Normans. Whey was a form of payment for farm workers in Ireland as well.

Food that is given to workers is always regarded as having low status, and by the sixteenth century, whey and buttermilk were not highly thought of in England. In 1615 the poet Gervase Markham recommended giving buttermilk away to poor people:

> The best use of buttermilk for the able housewife is charitably
> to bestow it on the poor neighbors whose wants do daily cry
> out for sustenance and no doubt that she shall find the profit
> thereof in a divine place as well as in her earthly business.

About whey, Markham wrote, "Its general use differeth not from that of buttermilk, for either you shall preserve it to bestow on the poor, because it is a good drink for the laboring man, or keep it to make curds out of it, or lastly to nourish and bring up your swine."

The Scots had a dish whose popularity lasted into the twentieth century. Cream and whey were beaten with a frothing stick, and after the mixture thickened, toasted oatmeal was sprinkled on top.

Ever since the Vikings settled Iceland in the ninth century, whey has been an important product there. This may also have been true even earlier, when the island was inhabited by Celtic monks. A rocky, volcanic, glacier-studded land, surrounded by rich fishing grounds but containing few trees, fruits, grains, grasses, or vegetables, Iceland

An Icelandic milkmaid, 1922, from People All Nations: Their Life Today and the Story of Their Past, volume IV: Georgia to Italy, *edited by JA Hammerton and published by the Educational Book Company, London, 1922.* (HIP/Art Resource, NY)

was not a place where anything edible was discarded. Its wide rocky swaths were suitable for sheep, but there was only limited grassland for grazing cows. Originally, most Icelandic dairy products were made from sheep's milk.

In Iceland's most famous novel, *Independent People*, by the Nobel Prize winner Halldór Laxness, Rosa, a farmer's wife, suggests getting a cow. This is enough to convince her husband, Bjartur, that Rosa is having a nervous breakdown. "Where's your field, then?" he demands. She says that she has found this one rich grassland. "When I'm busy in the meadow raking," she says, "I think about milk."

The reality was that there were few grasses hearty enough to live through the Icelandic winter. And the only cows that could survive were the brown-and-white ones introduced by the Vikings, producers of excellent milk from the island's tough but unusually rich grass. Even today, the country's brown-and-white cows—like its sheep, horses, and people—are direct descendants of Viking stock.

Skyr, originally a sheep's-milk product and later one made from cow's milk, is a by-product of a by-product. When Icelanders separated cream to make butter, they were left with skim milk. They soured it, and condensed it, to produce a unique product vaguely resembling yogurt. It is much denser than yogurt, however, and far more complicated and costly to make.

In making skyr, a great deal of whey is produced by condensing the moisture from the skim milk. And originally, the whey was not a by-product of making skyr, but the other way around—it was whey they wanted, and skyr was a by-product. Skyr was at first commonly eaten as a kind of porridge, cooked with Icelandic moss, because there was little grain in Iceland.

The name *skyr* is thought to have come from *skeroa*, meaning to cut, or *skilia*, meaning to divide. The making of skyr has been documented back to the fourteenth century, though earlier versions of it, referred to as "curd," may not have been as smooth as the ones produced now. In the National Museum of Iceland are dairy remnants found by nineteenth-century archaeologists that are thought to be skyr dating back to the year 1000. Skyr is also mentioned in the medieval Icelandic sagas, the starting block for both history and literature in Iceland. In *Egil's Saga*, which takes place in the ninth and tenth centuries but was written in the thirteenth, a scene unfolds in which Egil and his men are eating bowls of curds, thought to be skyr.

But in medieval times, it was whey, far more than skyr, that was in demand. Whey was often made into a drink, *mysa*. In the Króka-Refs saga, *The Saga of Ref the Sly*, set in the tenth century though written in the fourteenth, the king explains that "there is a drink in Iceland called *mysa*."

Because the Icelanders had no grain, they could not make beer like other Northern Europeans. And so the standard Icelandic beverage was soured whey. People fermented the whey by putting *mysa* in barrels with holes in the lid. For a time, impurities, such as bits of skyr, would bubble up through the holes and be removed. The barrel would then be retopped with fresh whey and sealed. The longer the whey was barreled, the sourer it became. Some drank it after a few months and others kept it for years.

The soured *mysa* turned into an alcoholic drink called *syra*. A small amount of *syra* was added to water to make *blanda*. Sometimes herbs or berries were added. Sometimes a bag of thyme was added to the whey barrel.

Syra was the standard drink of Iceland for many centuries. Before they went to sea, fishermen were guaranteed a supply of *syra*. When they couldn't get it, they tried to make a substitute by placing soured herbs in water.

Poor Iceland not only lacked grain, it also did not have much salt, which is why its commercial fish was air-dried rather than salted—stockfish, dried unsalted fish, rather than salt fish. Iceland did not even have enough strong sunlight to make sea salt. But whey could be used to preserve food. Butter was preserved in *syra*, which made it sour rather than salted. Blood sausages were also preserved in *syra*, as was a wide range of meats, fish, and vegetables. This made for a cuisine that was predominantly sour rather than salty. And in a stellar example of how the poor could make something out of nothing, sheep bones were left in *syra* to eventually decompose and be cooked into a sour high-calcium gruel. Sometimes, too, a little *syra* was added for flavor, the way other cultures use salt.

NUMEROUS CULTURES HAVE made cheese from whey, which is a triumph of conservation—making a valued product from what many thought was a worthless by-product to feed pigs. The most famous whey cheese is Italian ricotta, a name that expresses this economy of production, as it means "recooked." Milk is cooked to make cheese, and the whey that is pressed out is recooked at a very high temperature to make ricotta. There are cheese descriptions from ancient Greece that sound as if they could be ricotta, but it is generally thought that ricotta originated in medieval Sicily. The name for it there was *zammatàru*, which means "dairy farmer" and comes from the Arab word *za'ama*, which means "cow." Despite this origin, it is widely held in Italy today that the best ricotta is made from sheep's milk. Some historians posit that ricotta was developed between the ninth and eleventh centuries when Arabs occupied the island, but at the very least, Sicilian ricotta is as old as the eleventh century, when it appeared in a Latin translation of a book by the Arab doctor Ibn Butlan.

During the Renaissance, the food writer Platina described ricotta in this way:

It is white and not unpleasant to taste. Less healthful than fresh or medium aged cheese, it is considered better than aged or over salty. Cooks mix it into many vegetable ragouts.

Drinking whey has fallen out of fashion in most countries today, but it endured well into the nineteenth century, primarily for health reasons. Platina suggested drinking whey as a medicine "because it

Making ricotta, from Tacuinum Sanitatis, *originally* Taqwim es siha
*[The preservation of health], ca. 1445–1451 by Ibn Butlân, an Iraqi physician.
Painting on paper.* (Bibliotheque Nationale de France, Paris © BnF,
Dist. RMN-Grand Palais/Art Resource, NY)

cools the liver and blood and makes a way to purge the body of
poisons." An 1846 London book of Jewish cooking, *The Jewish
Manual*, authored anonymously by "a lady," recommends three whey
drinks—plain, wine, and tamarind, all of which appear in a chapter
titled "Receipts for Invalids." Apparently, at the time, a Londoner
who wanted to drink whey could no longer buy it but had to make it
from milk herself. Here are *The Jewish Manual*'s three whey recipes:

First, the plain recipe:

> Put into boiling milk so much lemon juice or vinegar that it will turn
> it, and make the milk clear, strain, add hot water and sweeten.

Next, one with wine:

> Set on the fire in a saucepan a pint of milk, when it boils, pour in as much white wine as will turn it to curds, boil it up, let the curds settle, strain off, and add a little boiling water, and sweeten to taste.

Finally, one that used tamarinds:

> Boil three ounces of tamarinds in two pints of milk, strain off the curds and let it cool. This is a very refreshing drink.

The Jewish Manual's chapter for invalids also offers recipes for numerous other milk dishes. Among them is "restorative milk," which is cooked with isinglass, a product from the dried swim bladders of fish that is usually used for clarifying gelatin. Another recipe, harking back to the ancients, is for making milk porridge:

> Make a fine gruel with new milk without adding any water. Strain it when sufficiently thick, and sweeten with white sugar. This is extremely nutritive and fattening.

The manual's author does not specify which grain to use when making porridge, though barley was much favored at the time. But how many recipes can be found for how to gain weight?

IN CORNWALL AND the west country of England, not all cream was used for making butter. Clotted cream was also a tradition. The earliest written records of clotted cream date from the sixteenth century, but the food is probably even older. Originally, it was developed as a way of preserving cream. Fresh cream turns very quickly; bottled clotted cream can keep for two weeks.

When making clotted cream, as when making butter, fresh milk is left for about eight hours until the cream rises to the top. Today, this is done by separator machines, but in earlier times, milk from a morning milking would be left out until late afternoon and milk from an afternoon milking would be left until morning. Then the milk would be slowly simmered for several hours at a low heat in a shallow brass or earthenware pan over a low charcoal fire until the cream had a bubbly crust. It would then be left to cool, and the clotted cream, which is much thicker than natural cream, would be skimmed off.

In much of Europe, there was also a tradition of freshly curdled milk—which the Romans had recommended as the most healthful milk. In Cornwall it was called junket. Hannah Glasse, in her 1742 *The Compleat Confectioner: or The Whole Art of Confectionary Made Plain and Easy*, gave this recipe for junket:

> Take a quart of new milk and a pint of cream; put it warm together, with a spoonful of good rennet, and cover it with a cloth wrung out of cold water; gather your curd, put it in rushes till the whey is run out and serve it either with or without cream.

Note the direction for using "new milk." Experience had taught cooks that when working with fresh milk that needs to stand, the fresher the milk the better.

Hannah Glasse also offered this pleasant variation of junket, which she called "stone cream":

> Take a pint and a half of thick cream, boil it in a blade of mace and a stick of cinnamon, with six spoonfuls of orange flower water, sweeten it to your taste, and boil it till thick, pour it out, and keep stirring till almost cold, then put in a small spoonful of rennet, and put it in your cups or glasses, make it three or four hours before you use it.

The tradition of day-old cheese being transformed into a dessert—which is essentially what happens with junket—appears in many older European cultures as well. The Scots also made junket, which they called *hattit kit,* or "hatted kit." This is not a reference to haberdashery, but an obsolete word for coagulation. *Hattit kit* is always made with buttermilk, and the original version was made on the farm. A pan of buttermilk was brought to the cow and warm new milk squirted in it to start the buttermilk curdling. Later on, rennet was used instead of buttermilk and thick cream instead of warm new milk.

In this 1977 recipe from the Scottish Women's Rural Institutes, both techniques were combined. The cooks also pointed out that although the milk did not really have to be milked directly from the cow into the pan, that was still the best way to do it.

> Warm slightly over fire 2 pints of buttermilk. Pour it into a dish and carry it to the side of a cow. Milk it into about 1 pint of milk, having previously put into the dish sufficient rennet for the whole. After allowing it to stand for a while, lift the curd, place it on a sieve, and press the whey through until the curd is quite stiff. Season with sugar and nutmeg before serving, whip some thick cream, season it also with a little grated nutmeg and sugar and mix gently with the curd.

In the Basque language, desserts made from cheese are called *mamia.* Or, in Spanish, *cuajada,* meaning curds. The Basque made these desserts with sheep's milk because the Basques made everything with sheep's milk. Their use of cows didn't begin until modern times, and though cow's milk is now sometimes used in making cheese, *mamia* is still made exclusively from sheep's milk.

The Basques heat the sheep's milk with a small pinch of salt and bring it to a boil. Then the milk is removed from the heat, and when its temperature drops to about 85 degrees Fahrenheit, a teaspoon of rennet is added. Next it is stirred vigorously with a spatula and poured into cups. Traditionally, wooden cups were used, but today, small earthenware cups are preferred. It is left in a cool place to set (refrigeration can speed up the process), and after a day, it is ready to eat. Honey is poured on top. It is fresh and mild-tasting, with the sweet flavor of sheep's milk.

If *mamia* is not eaten after a day or so, it can serve as the beginnings of a sheep cheese. In Catalonia a fresh cheese similar to *mamia*, called *mató*, is made and served with honey. Though often produced with cow's milk today, it was traditionally made from goat's milk.

In sixteenth-century Britain, possets and syllabubs, made in a similar vein, though without rennet, were popular. Both milk-based, possets were served hot in decorative metal or ceramic pots and syllabubs were served cold in decorative glasses. Syllabubs were probably invented in Tudor England but, like the Tudors, made their way into Scotland, where they enjoyed some popularity. Possets are older; they seem to date back to the Middle Ages as a form of curdled milk. Later, fruit peel and sherry were added to possets.

Elizabeth Cleland's 1755 recipe for syllabub, like the traditional *hattit kit* recipe, calls for a visit to the milking shed. Her recipe also shows the belief that fresh milk should be consumed as quickly as possible. Warm milk taken directly from a cow is very frothy and may be the origin of the frothy top of both dishes.

TO MAKE SILLABUBS FROM THE COW

Sweeten either wine, Cedar [cider] or strong ale, put it in a bowl, take it to the cow, and milk her on your liquor as fast as you can. You may make it at home by warming it, and pour it on the liquor out of a teapot.

Cleland also offers two more usual recipes:

SOLID SILLABUBS

Take a chopin of very thick cream, put into it three gills of Malaga [sweet sherrylike fortified wine], the grate of a lemon, the juice of two bitter oranges, and sweeten it to your taste. Beat it well for a quarter of an hour, then skim it with a spoon, and put it in glasses.

A SACK POSSET OR WHAT IS CALLED THE SNOW POSSET

Boil a chopin of cream or milk with cinnamon and nutmeg; then beat the yolks of ten eggs. And mix them with a little cold milk; then by degrees mix them with the cream, stir it on the fire until it is scalding hot; sweeten it to your taste, put in your dish a mutchkin of sack [sherry] with some sugar and nutmeg; set it on a pot of boiling water, and when the wine is hot, let one take the cream, and another the whites of the eggs, and pour them both in holding your hands high, and stirring all together while it is on the fire; when it is scalding hot, take off, cover it, and let it stand a while before you send it to the table. The whites must be beaten with a little sack.

Though Cleland is not clear on this point, the whites are supposed to be a froth added to the top of the posset.

Another idea that became very popular in the eighteenth century was dessert creams, which were nothing more than flavored, thickened, whipped creams. Cleland gives recipes for eighteen creams of various degrees of complexity. One of the most popular was steeple cream, so known because it was formed into a cone shape resembling a steeple.

5

DESERT MILK

Wᴴᴇɴ ᴍᴏᴅᴇʀɴ ᴘᴇᴏᴘʟᴇ think of a milk culture they think of
America today or of early northern Europeans. But there has
never been a culture more dependent on milk than the desert nomads
known as the Bedouins. Today the Bedouins are losing their traditional way of life—less than 10 percent are still nomadic. But for
centuries, the Bedouins were nomads who, unlike other Arabs, lived
out in the desert. The word "Bedouin" means "desert dwellers."
Among the first to adopt the religion of the Prophet Muhammad,
the Bedouins were devout Sunni Muslims and were much admired
by the Arab world. And yet they built no mosques. Or anything else.
The Bedouins didn't build things. They prayed facing Mecca in the
open desert, and in ceremonies that involved washing, they washed
with sand. They spoke their own language, had their own customs,
belonged to no country, and moved continually.

At times, the Bedouin diet consisted almost entirely of milk. The
milk came from camels, their most important possession. The
Bedouins lived with camels, pitched tents made of camel hair, slept
next to camels, traveled by camel, and milked camels. All over the
world, nomadic herders lived on milk, but camels are the rare large
mammal that is suited to desert life. They eat food that a human
would not recognize as food. Sauntering through a desert area of
rock and sand that seems barren, they suddenly leave the trail to
stick their head around a rock and find a camel-thorn salt bush
to munch on.

The camels' diet does not seem ideal for milk production. If, for example, they feed on the spiky salt bush, more correctly called *atriplex*, whose roots soak in salt, their milk has a very salty taste. But the Bedouins were, and are, willing to drink salty milk because camels need salty plants. When their salt intake is lowered, they produce less milk. Other unlikely fodder leads to other odd tastes. Nonetheless, the camel, seemingly out of nothing, produces a protein- and fat-rich milk. And, in times when there is a pronounced lack of water, the camel has the ability to dilute her milk to provide her young, and by extension her herder family, with more water. So, depending upon environmental circumstances, camel's milk varies greatly; at times, it has far more fat and protein than goat, sheep, or cow's milk, but at other times, it has far less.

The Bedouins usually covered their camels' udders with cloth to keep the young camels away and reserve the mothers' milk for human use. They and other desert Arabs milked into a bowl that rested on one knee, with a foot on the other knee; it was an unnatural balancing act, but the camel has an unusually high udder. Some Arabs kept the milk in goatskin or camelskin bags and made it into *laban*, yogurt, but the Bedouins, constantly traveling, drank it fresh, and preferred it direct from the udder. Because it was warm, they called this milk "cooked" milk.

★   ★   ★

IN THE SEVENTH century, the Arabs, the people of the Arabian Peninsula, spread as far north as Iraq and Syria, as far west as the Atlantic coast of Morocco and Spain, and as far east as Persia (Iran). The Arab Empire became one of the largest empires in history, and the Arabs, the ruling elite, reigned over populations numbering larger than their own, populations with cultures, languages, and traditions different from the Arabs'. The one common thread tying the empire together was Islam, both the religion and the culture.

MUHAMMAD AND THE Qur'an had very specific instructions regarding breastfeeding. Babies were to be breastfed until the age of two. Thus, in the early years of the Arab Empire, few babies were bottle-fed. The edict eventually became contested, however, because a lengthy lactation period is a form of birth control and some religious leaders wanted to encourage large families.

Parents were assured that if an infant died before the two-year nursing period was up, the child would be breastfed in Paradise. Believers were assured that Muhammad was breastfed after an easy birth, and this did not seem surprising, because it was also said that he was born already circumcised. However, a dissident view in the late fifteenth century claimed that Muhammad's birth had been very difficult and his mother had been able to breastfeed him for only a few months. This raised the question of whether he had been given to a wet nurse.

Islamic law did allow for wet-nursing, and believers ascribed to the ancient, persistent belief that the traits of a wet nurse could be passed to the child. Muhammad forbade the use of wet nurses who showed signs of mental unbalance. Avicenna, a highly influential tenth-century doctor in the Persian court, had a great deal to say about the prerequisite qualities of a wet nurse. He also allowed that if a wet nurse was physically or mentally indisposed, a baby could be artificially fed until she was cured. But artificial feeding was rare in the medieval Muslim world. And few bottles or other feeding vessels have been found.

Maimonides, the famous twelfth-century Jewish doctor and theologian from Córdoba, appears to have been influenced by his Muslim teacher, Averroes. In his *Book of Women*, Maimonides writes that mothers are to breastfeed for two years, during which time they are to abstain from sex. He also specified that if a woman had twins, she was to breastfeed one and hire a wet nurse for the other. Babies were not to be fed animal milk.

MANY DAIRY PRODUCTS were used for cooking in the Arab Empire, including fresh milk, soured milk, and cheese. Among them was *liba*, which the ancient Greeks called *pyriate*, a product made from colostrum, the first milk produced in lactation. Yellowish and sticky, it does not look or feel like milk. Used as food for a newborn, it contains a high concentration of antibodies, protein, white blood cells, vitamins, and zinc, and it is very low in calories because it has little fat and lactose. It helps protect the newborn from diseases and acquire important building blocks even as the infant loses a slight amount of weight. Lactating mothers produce colostrum for only about three days. In modern times in the West, it became known as "liquid gold."

When making *liba*, the Arabs mixed milk and colostrum together, sometimes in equal parts and sometimes using two parts milk. The fourteenth-century cookbook *Kitab Zahr al-Hadiqafi al-At'ima al Aniqa* said that peasants sometimes made liba entirely with colostrum, but that it had an unpleasant flavor until they learned to mix it with milk. The *liba* was mixed and cooked and left outside on a warm night. By morning it was solid. And for those who could not get real colostrum, which was available only right after birth and often had to be saved for newborns, there was a fake *liba* made with egg whites and one yolk and cooked into what was essentially a sugarless custard.

Biráf was milk that was left in the warm night air so that it had soured by morning. *Kitab Zahr al-Hadiqafi al-At'ima al Aniqa* said that it could be eaten plain or with honey, syrup, or sugar. It added that doctors recommended sucking on a quince afterward or sipping a drink made with vinegar and quince-flavored syrup, what in contemporary America is known as a quince shrub.

The Arabs ate a cheese called *halum*. Goat's or sheep's milk was boiled with thyme until reduced by about a third. Then it was cooled, mixed with rennet, and layered in a mold with fresh thyme and peeled citrus. Next, boiled milk was poured over it, and it was covered with olive oil to seal out air. *Halum*, which is a stretched cheese like string cheese, is still made today, but with mint rather than thyme.

Yogurt was commonplace, and sometimes it was drained to produce a denser cream called *qanbaris*. A kind of soft cheese called *shiraz* was made by adding rennet to yogurt. Sometimes it was mixed with barley or wheat, making a porridge.

IN THE TWENTIETH century, Spanish scholars found an anonymous manuscript of recipes from thirteenth-century Arab Spain and Morocco that confirmed the Arab use of dairy products and contained the following recipe:

RAFIS WITH SOFT CHEESE

They take crumbs of clean bread, when removed from the oven, and knead the crumbs only without the crust and they are soaked with another of fresh soft cheese without salt and with some butter; it is shaped and melted and clarified butter is poured over it, and the desired amount of cleaned honey without foam.

Note the use of clarified butter, known in Arabic as *smen*, although judging from its infrequent citations in this manuscript, the word was not in common usage in the Middle Ages. *Smen* is still used throughout North Africa and in parts of the Middle East. It is also widely used in India, where it is called *ghee*. Clarified butter is an intelligent way of dealing with butter in a hot country because once the fat solids are removed, the butter becomes a clear oil with a very long shelf life. It is often aged by burying. In southern Morocco, the Berbers, the original pre-Arab inhabitants of North Africa, seal and bury jars of *smen* at the birth of a daughter and then dig them up for use at her wedding. In Morocco, *smen* is often made from sheep's milk, but cow's milk is considered better.

It is often claimed, though it is probably not true, that the second-oldest book on Moroccan cuisine after the thirteenth-century manuscript found by the Spanish scholars is a small book by Zette Guinaudeau published in 1958. The wife of a doctor, she wrote on the cooking of Fez. This was her guide to making *smen*:

> When melted, the butter is skimmed during boiling, which lasts about a quarter of an hour, and then strained through a fine cloth into a *khabia* [crock], where it is kept. Salt it lightly before it curdles, stirring with a wooden spoon. A white deposit remains at the bottom of the cooking pan. Use a ladle for pouring the butter (the solids remain on the bottom).

Arabs also have a tradition, dating back many centuries, of drinking *lben*, which is soured whey left over from making curds or cheese. Madame Guinaudeau, as she was always known, turned lyrical on the subject of *lben*:

> Lben! . . . Lben! The street cry is heard in Fez as soon as the fine weather arrives. Drunk by rich and poor, city dweller and "fellah," it is offered in the shops to all who pass by. May you one day know the joy of quenching your thirst with this slightly acidic whey, drunk in the pale shadow of an olive tree after a long excursion on a hot day in June.

In 1326, Ibn Battuta, from an educated family of religious authorities, set out from his native Tangiers on the far western end of
the Arab Empire and went on a pilgrimage to Mecca, as all devout
Muslims are supposed to do at least once in their lifetime. But he
didn't stop with Mecca, and over the next twenty-seven years he
traveled more than 75,000 miles, to visit every land under Muslim
rule.

In his travels Battuta often encountered milk, usually soured milk,
because fresh milk was a rarity. In Abyssinia, northern contemporary Ethiopia, he was offered plantains boiled in fresh milk, which
curdled the milk, and then the plantains and the curds were served
in two separate dishes, apparently an unusual enough meal for him
to make note of it in his diary. In Mali, when the sultan said he was
sending him a gift, Battuta imagined rare textiles, fine clothes, a
horse—or "horses," as he wrote in his diary—but when the gift
arrived it was plates of food, "three of bread, with a piece of fried
fish, and a dish of sour milk." Apparently milk and fish were far more
rare and valuable in Mali than in Tangiers. Battuta noted, "I smiled
at their simplicity, and the value they placed on such trifles as these."

One of the best sources on medieval Arab cooking is a Baghdad
manuscript from the year 1226, when the city was still an important
political and cultural center. Thirty years later it would be destroyed
by the Mongols. The author of the cookbook was Muhammad Ibn
al-Hasan Ibn Muhammad Ibn al-Karim al-Katib al-Baghdadi.
Despite all the information embedded in that name, we know little
about him other than the fact that he was from Baghdad.

From this Baghdad cookbook, it is clear that while the Muslims
borrowed a great deal of dietary restrictions from the Jews, they
never were concerned with the Jewish interdiction on mixing meat
and dairy. Cooking meat with yogurt was common in the Muslim
world. Also, in these recipes, yogurt is always referred to as Persian
milk, showing that the Arabs, or at least the Baghdad Arabs of the
Middle Ages, considered yogurt an idea that had come from Persia.

Most historians think that this early yogurt was sour yogurt, and some think that it was thickened with rennet.

Here is one recipe from the Baghdad cookbook:

MADIRA [THIS NAME COMES FROM *MADIR*, WHICH MEANS "CURDLED"]

Cut fat meat into middle sized pieces with the tail. If chickens are used quarter them. Put into the saucepan with a little salt and cover with water: boil removing the scum. When almost cooked, take large onions and Nabatean leeks, peel, cut the tails, wash in salt and water, dry and put in the pot. Add dry coriander, cumin, mastic [dried sap from the mastic tree with piney aroma common in Arab cooking], and cinnamon, ground fine. When cooked and the juices are dried up, so that only oil remains, ladle out into a large bowl. Now take Persian milk as required, and put into the saucepan, add salted lemon [lemons preserved in brine] and fresh mint. Leave to boil, then remove from the fire, stirring. When the boiling has subsided, put back the meat and herbs. Cover the saucepan, wipe its sides, and leave to settle over the fire: then remove.

Evidence that yogurt came to the Arab world from Persia can be seen in the Persian *borani*, a popular yogurt dish, frequently made with eggplant, that in time spread to Iraq, Afghanistan, Pakistan, Armenia, and Georgia. The dish was named for the Sassanid queen Boran of the ninth century, who was said to love yogurt. She was also said to love her husband, the caliph al-Ma'mun, and traveled with him on military campaigns. She was known for her sense of style. Great palaces and mosques were built both by her and in her honor. But none have survived, only her beloved yogurt and eggplant dish. *Borani*s are often confused with the completely unrelated Indian dishes *biryani*, which comes from another Persian word and means "fried."

The earliest surviving recipe that resembles modern eggplant *borani* comes from the fourteenth-century *Kitab Wasf al-Atima al-mu'tada* [*The Description of Familiar Food*], which may have been printed in Cairo but was based on Baghdad cooking:

> Fry eggplant in sesame oil or fresh tail fat and peel them and put them in an ample vessel. Then pound them with the ladle until they become like pounded *harisa*. Then you throw on Persian yogurt in which you have garlic pounded with a little salt and mixed it well with it. Then take pounded lean meat and make meatballs from it and throw them in the tail fat and put them on the surface of the eggplant and yogurt. You sprinkle dry coriander and Chinese cinnamon on it, both pounded fine and it comes out well.

Yogurt and vegetables, yogurt and meat, and yogurt and fish dishes were popular. This chard and yogurt dish is from *The Description of Familiar Food*. The recipe says to boil the chard with salt, but the usual Arab practice was to boil green vegetables with natron—a naturally occurring bicarbonate of soda, i.e., baking soda, found in the desert—combined with other salts. This kept the natural colors bright.

SILO BI-LABAN. CHARD AND YOGURT.

Take big Swiss chard stalks and cut off their leaves and throw them away. Cut them the size of four fingers and wash them. Then boil them in salted water until done. Take them out of the water and spread them out on a woven tray until dried off. Put them in yogurt, which has been mixed with garlic, which has been boiled, and spread a little nigella [a black seed of an Asian flowering plant often called black cumin, though it is not related. Today its best-known use is in Armenian string cheese] and mint leaves on it.

Salt fish and yogurt, *Samak malih bi-laban*, from the same book, is a perfect desert meal:

> Take salted fish, wash it, cut it up medium and fry it. Then take it from the pan hot and put it in yogurt and garlic. You throw nigella

and finely milled Chinese cinnamon on its surface, and it is eaten hot or cold.

Another use of yogurt popular in the Arab empire was *kamakh rijal*, which was yogurt and salt left to bake in the sun. This was often an urban dish made on household rooftops. It ages like cheese and after about a month, even develops a cheeselike smell. Protected from decay by its salt, it continues aging for months. Fresh milk is mixed in daily, and it ends up resembling a soft salty cheese. It has a sharp flavor, especially when made, as recommended, with sheep's milk.

Arabs did also cook with fresh milk, often together with rice. This recipe is from the thirteenth-century Baghdad cookbook:

RUKAMIYAH [ROUGHLY TRANSLATED AS "WHITENESS"]
Cook rice with milk until set thick, then ladle out. Place on top of this, meat fried in tail-fat and seasoning in the form of kebabs. Sprinkle with cinnamon.

Arab dishes with fresh milk as the central ingredient were rare, but there were a few, such as this one from *The Description of Familiar Food* called *rukhamiyya*, or "marbling." It uses sugar; the Arabs were the first to use sugar.

A quart of rice, three pounds of milk, a race [root] of ginger, a stick of Ceylon cinnamon, the weight of a quarter dirham [the monetary unit] of mastic. Then you take half of that milk—that is, a pound and a half—and you put it in the pot, and you put the Ceylon cinnamon, ginger, and mastic with it. When the water and the bulk boil ["water" may be a mistake, since there is no other mention of it] you wash the rice and put it in the pot, moistening it with the pound and a half of milk that remains, little by little, on a gentle fire. Then you moisten it awhile, and you stir it awhile, and you continue moistening it and stirring it little by little on a nice fire. If all that is on a fire of coals, best of all. And when it smells good you leave it hanging overnight in smoke. And when something of the dish tastes smoky to you, take a bunch of leeks, cut off their ends and tie them and arrange them in the pot while it is hanging in the smoke. When you ladle it out, ladle it with sesame oil and sprinkle it with sugar.

6

THE DAYS OF MILK AND BEER

MEDIEVAL EUROPEANS COUNTED on dairy products—they were central to their diet—especially cheese. Milk was also used, occasionally for drinking but much more commonly in cooking.

In the fourteenth century, Guillaume Tirel, more often known as Taillevent, the chief cook for Charles V of France, wrote a cookbook that is considered to be the founding document of French *haute cuisine*, French cooking for the upper class. Milk is not a leading ingredient in his recipes, but it does show up from time to time, often in unexpected ways.

Taillevent avoided mixing milk with fish. This was a medieval taboo, though it was not always strictly adhered to. Jews were more concerned about mixing meat and dairy. The Dutch, irrepressible dairy consumers, ate herring with sour cream. But as far as we know, Taillevent and his royal patrons never combined fish and dairy, and in medieval Europe in general there are few recipes to be found with this combination. An uneasiness about it persists, though there is little reason for it. In Italy today it is a major gastronomic faux pas to put grated cheese on pasta that contains seafood.

Taillevent made a dish made with milk curds and *lardons*, which are pieces of smoked pork breast, as opposed to American bacon, which comes from the belly. The dish was not a Taillevent original; a similar dish was found in an earlier German manuscript. But this recipe is unusual in that it offers the cook the option of using fish instead of bacon, thus breaking the fish-dairy taboo. When the dish was made without meat, it could be served on holy days.

LARDED MILK

Boil milk on the fire, beat some egg yolks, and put the milk over a fire with a small amount of charcoal [low heat] and add the eggs. Those who want meat [i.e., if it is not a holy day], cut two or three pieces of *lardon* and put it with the milk, or if you want to add fish, there is no need to use *lardons*, but add wine and verjus [the very tart juice of unripe grapes, popular in medieval cooking] for the purpose of curdling, then remove from the fire and drain it on a white cloth and fold it and squeeze it. Put it on a table to drain for three days with some clove buds and fry until it gets some color and toss sugar on top.

Taillevent also gave this milk recipe:

PROVENÇAL MILK

In a sauce pan scald cow milk. Beat egg yolks off the fire and while doing this slowly add about a quarter cup of hot milk into the eggs and stir the mixture into the remaining milk. Cook this until the sauce is thick. Poach eggs in water and gently add them to sauce. Put them in bowls and serve with toast.

The influence of Taillevent's recipes can be seen in later works such as the 1393 guide for bourgeois households, *Le Ménagier de Paris*. Some of these variations of his recipes could be seen as improvements. For example, *Le Ménagier* added ginger and saffron to the Provençal milk recipe.

Fourteenth-century France also had a kind of milk sauce for poultry. Called *dodine*, it was made from the drippings of a roast. Like porridge before it and pudding after it, *dodines* were not originally made with milk. This *dodine* for capon and lamprey, by Chiquart, a

Savoyard court chef in 1420, uses cheese. The recipe also shows that Brie, the famous soft cheese dating back to at least the eighth century, was a well-established food by the fifteenth century. The Crampone cheese mentioned comes from the Auvergne and is similar to Gruyère:

> When your capons are off the spit, get a good pot that is very clean and get a good strainer and strain into the pot what you have collected in the silver dish or pan from the capon and lampreys, and if you see that there is not enough broth to make the Dodine, draw it out with your good beef bouillon. Get white ginger and a few grains of paradise [sometimes called melegueta pepper, a peppery seed] and flavor it with verjuice, though not too much, and with salt, too. Get good parsley, chop off its leaves; get your table bread and cut good slices for toasting; get very good Crampone cheese or Brie cheese, or the best cheese that can be had; of the slices of toast cut each into three strips, then set them out on your dishes, with cheese over them, and then your broth over top of that. When they go to the dresser those sops are served up [that is, the slices of bread with sauce and cheese are put on a plate] and in another dish put the Pilgrim Capons.

An even earlier *dodine* from *Le Grand Cuisinier de Toute Cuisine* by an unknown author in the mid-fourteenth century takes a yet bolder turn:

WHITE DODINE

Take some good cow's milk. Put it to cook in a frying pan beneath your roast with some white powder [sweet spices], two or three egg yolks: strain your milk and let it cook all together with a little sugar some salt and a little bit of parsley leaves. If you want, put in some minced marjoram. Place your roasted waterfowl on top.

Something was beginning here—an early use of sugar in Europe. Before this, there were a few sweet dishes made with honey, but desserts in the form of sweet dishes served at the end of the meal were not a widespread practice.

In fourteenth-century France and England, there was a growing tendency to combine milk with sugar. By the sixteenth century, sugar, which had been a luxury for the rich, started to become more affordable and available. This was especially true after 1493, when Columbus brought sugar to the Caribbean and unleashed one of the greatest crimes of human history: the African slave trade, which made sugar inexpensive.

A pie was often a savory, not a sweet, dish. The unidentified A.W. who wrote *A Book of Cookrye*, originally published in London in 1584, offered recipes for mutton, chicken, calf's-foot, and veal pies. Even his cheese tart was unsweetened. But by then, cheese tarts were a very old recipe. Taillevent had offered one more than two centuries earlier. This is A.W.'s recipe:

TO MAKE A TARTE OF CHEESE

Take good fine paste and drive [roll] it as thin as you can. Then take cheese, pare it, mince it, and braye [mash] it in a morter with the yolks of Egs til it be like paste, then put it in a faire dish with clarified butter, and then put it abroade into your paste and cover it with a faire cut cover, and so bake it: that doon, serve it forth.

The unidentified A.W. also gave recipes for fruit pies—quince, apple, strawberry, and more. Fruit was one of the first desserts, followed by baked and sugared fruit. Sweet pies were well under way in the sixteenth century.

A.W. also gave a recipe for a sweetened cream tart. This was a newer kind of cooking. There had been sweetened dairy before— such as honey and yogurt foods and syllabubs—but not dairy desserts of this kind: pies, puddings, custards, and creams made from sugar and eggs. Here is A.W.'s cream tart:

TO MAKE A TART OF CREAM

Take Creame and Egs and stir them, together, and put them into a strainer till the whey be come out, then strain it that it may be thick, season it with Ginger, Sugar, and a little Saffron, and then make your paste with flower [sic], and dry your paste in the Oven, and then fill it, and set it into the Oven to dry, and then take it out, and cast Sugar on it, and so serve it forth.

Medieval Europeans were always cautious about milk. The debate over which animals produced the most healthful milk had continued. Taillevent often specified using cow's milk, while *Le Ménagier de Paris* cautioned that the ill and convalescent should avoid it. The best milk, it asserted, was human milk, followed by donkey's milk, sheep's milk, and goat's milk. But the Normans, who ruled England after 1066, did not like goat's milk.

Curiously, although most Europeans rated donkey's milk very highly, they had no interest in mare's milk, which has similar qualities. Europeans had horses, but left their colts to suckle on their mares. They knew that Asians used mare's milk. Homer had mentioned it, calling Scythians "mare milkers," and Herodotus wrote of the custom of drinking it in Central Asia. But, as often happened when he wrote about foreign customs, he did not make it sound good. It starts well, when he describes them skimming the cream off mare's milk to enjoy it as a particular treat. But then he writes about how a blind slave would stick a blowpipe made from a hollowed bone into a mare's genitalia and blow to expand the udder, while another slave did the milking.

Early Europeans who traveled to Mongolia had nothing good to say about the culture and milk there. They considered the Mongols the epitome of barbarism. In 1246, Friar John of Pian de Carpine visited the Mongol court and called the Mongols a detestable nation of Satan. He described their barbarous slaughter of people, villages, and gardens as they went "swarming over the earth like locusts" and said they drank blood. He even found their large horses barbaric. As for prisoners, he said, they were enslaved and horribly mistreated. He wrote, "They have misused their captives as they have misused their mares." If Mongols drank mare's milk, then drinking mare's milk was barbaric.

William of Rubruck, a Franciscan from Florence who traveled to Mongolia in 1254–1256, also had nothing good to say about the Mongols. He described an encampment of chieftains who kept their horses near their tents. The chiefs had a long meeting, "and there they remained until noon, when they began drinking mare's milk till evening so plentifully that it was a rare sight." Even those Europeans who regarded dairy as central to their diet would never spend an entire day guzzling milk. That scale of milk consumption was unimaginable to a European.

Rubruck quickly realized that the Mongols lived in a milk-based society of a level he had never seen before. Felt representations of udders hung over the entranceways of their tents. The mares were large and often uncooperative and had to be milked five or more times a day to produce only about two liters per horse—a very small quantity for so large an animal. He described the milking of the mares thus:

> They stretch a long rope on the ground fixed to two stakes
> stuck in the ground, and to this rope they tie toward the third
> hour to the rope the colts of the mares they want to milk. Then
> the mothers stand near their foals and allow themselves to be
> quietly milked, and if one be too wild, then a man takes the
> colt and brings it to her, allowing it to suck a little, then he
> takes it away and the milker takes its place.

Because the mares were so difficult to milk, milking was men's work, in contrast to most cultures, where women milked the cows.

William of Rubruck also witnessed how the Mongols made powdered milk. He wrote that they "boiled [the milk] down until it was perfectly dry and would keep for very long times without salting."

Fresh mare's milk is a strong laxative and is generally regarded as undrinkable. Rubruck described how some of this milk was made into butter by putting it in a skin and churning with a hollow stick that he said was the size of a human head on one end. The Mongols drank the tart liquid that was left over, which must have been buttermilk.

Rubruck also described two products whose manufacture is based on the fact that mare's milk does not curdle. This, too, would be its disadvantage to Europeans, because you cannot make cheese from it. But that is also true of donkey's milk, which Europeans favored, though they made very little. The Mongols would churn the mare's milk until its thicker part sank to the bottom. The bottom part was very white and was given to slaves. The top part was a clear liquid— whey—and was given to lords. The clear whey drink was fermented, and both Rubruck and Marco Polo drank it, enjoying the slight intoxication that resulted, though Rubruck said he had to first get over the shock of its tartness. It made him break out in a sweat, and it "brings on urination," he wrote.

The intoxicating drink, known as *koumiss*, was kept in a huge skin bucket in a family's tent. Each time someone entered, they gave a stir to the *koumiss* to help with fermentation. Rubruck also witnessed the Mongols' annual festival of May 9, which celebrates the first *koumiss* of the new year. A few drops of *koumiss* are sprinkled on the ground and all the white horses in the herds blessed.

Koumiss was a basic part of the Mongol diet, especially after the late sixteenth century, when the Mongols converted to Tibetan Buddhism and the eating of fish and horsemeat was banned. The success of the Mongol conquest of most of Asia and Eastern Europe, resulting in the largest empire in history until the twentieth-century British Empire, was basically achieved by mounted soldiers who could ride for long hours without stopping to eat or rest. They drew their sustenance from the *koumiss*, dried cow's-milk curds, and powdered milk that they carried.

Powdered milk, a common commodity today, was a curiosity for thirteenth-century Europeans. Marco Polo wrote:

> They also have milk dried into a kind of paste to carry with them; and when they need food they put it in water and heat it up until it dissolves and then they drink it. It is prepared in this way; they boil the water and when the rich part floats on the top, they skim it into another vessel, and of that they make butter, for the milk will not become solid until this is removed. Then they put the milk in the sun to dry. And when they go on an expedition, every man takes some ten pounds of this dried milk with him, and of a morning he will take half a pound of it and put it in his leather bottle, with as much water as he pleases. So as he rides along, the milk paste and water get well churned together into a kind of pap [that word was later used for milk-based baby formula] and that makes his dinner.

All Mongol males, no matter how high their rank, were required to give mares to the emperor so that he could provide himself and his family with dairy products. It was a horse culture back then and it still is today. Many men work and travel on horseback. Even signalmen holding up their batons beside train tracks are mounted. And they still drink *koumiss*.

At the end of the thirteenth century, after William of Rubruck but before Ibn Battuta, Marco Polo visited the Mongols. He wrote about Mongols drinking *koumiss*, about Kublai Khan's passion for it, and about how the ruler maintained a special herd of white mares as a private milk source. In fact, Marco Polo had a lot to say about milk, which is interesting, because despite his reputation for introducing many European food trends, he was really not much of a food writer. He completely missed the existence of tea, which later became so popular in England and Europe, and the well-known story of his introduction of pasta to Italy is completely apocryphal— as can be seen by the early Roman *tracta* recipes and the early Italian and Arab pasta recipes. But, like William of Rubruck, he seemed taken with *koumiss*.

Ibn Battuta made out better with the Mongols than he did later in Mali. Their chief sent him several sheep, a horse, and a large bag of *koumiss*.

WITH GOOD REASON, Europeans were uneasy about drinking fresh milk. A Catalan health code written in 1307 in Montpellier, *Regiment de Sanitat*, Rules for Health, by Arnau de Vilanova, a widely recognized medieval doctor, warned about milk, "You must consume it immediately after producing it and it should not have a strange taste or a bad smell. Then it is all right to drink a quarter of a liter." The twelfth-century English writer Alexander Neckam wrote, "Raw cream undecocted eaten with strawberries . . . is a rural man's banquet. I have known such banquets have put men in jeopardy of their lives." It was often advised that after drinking milk, the mouth should be flushed with honey to be rid of the harmful effects of the milk. It could be seen that people became sick and even died from milk even though no one understood why. Many would not take the risk of fresh milk, but indeed it is surprising that anyone would.

Milk was a bodily fluid, like blood, many believing it *was* blood, and in the Catholic Church it was under the same interdictions as red meat. No dairy product could be consumed on holy days. The number of these days steadily rose until the seventh century and then remained in place for the remainder of the Middle Ages. These days included Wednesday, Friday, and Saturday, and the forty days before Easter. In all, for considerably more than half the days of the year, dairy, along with red meat, could not be eaten.

In 1500 in the Norman city of Rouen, where one of France's three great cathedrals stands, there was a shortage of cooking oil. So the archbishop gave his dioceses permission to use butter. Each local diocese that wanted to take advantage of the special dispensation had to pay a small fee to the church, and legend has it that one of the towers of the cathedral was built with this money. The so-called butter tower, erected in the early sixteenth century, was a late addition to the church, which was begun in the twelfth century.

ONE OF THE few milk controversies that have vanished over time was whether milk was hot or cold. This was a subject of much debate in the Middle Ages, and it had nothing to do with temperature. Rather, it was a concept—and one that is still adhered to in China today. Some foods were regarded as "hot," and some as "cold," and since the body reacted differently to the two, good health depended on having an appropriate balance between them. Hot food caused sexual desire; red meat was a "hot" food, and this was the original reason why it was banned on holy days.

Which foods were hot and which were cold was generally straightforward. Red meat was hot; fish and anything that lived in the water was cold. Thus, whale meat was considered a cold red meat. A beaver was hot, but its tail was cold, as a beaver sits on a log with its tail in the water. And so dishes made of beaver tail or whale meat were appropriate on holy days, which led to the near total destruction of European whale pods.

Milk was a more complicated matter. Some thought it was hot, others found it cold, and still others classified it in a special category, "warm." For many who regarded milk as whitened blood, milk was simply hot, like red blood. But some, including Galen, argued that when blood turned into milk, it lost its heat. Galen even said that it was dangerously cold. Those who argued that milk was warm complicated matters by saying that the various products made from it might be hot or cold. Advocates of milk being warm or hot also argued that the very young and very old are cold and therefore milk was good for them, whereas for the average adult, milk was risky. And for those who were weak or sick or even melancholy, milk was ill-advised.

Nor was animal milk recommended for babies in medieval Europe. As in the ancient world, it was recognized that some women couldn't or wouldn't breastfeed, and for them the remedy almost always was wet nurses. A twelfth-century English text said, "The women of our own race quickly wean even those who they love, and if they are of the richer classes they actually scorn to suckle them." An English text by Bernard of Gordon in the fourteenth century recommended wet nurses because "Women nowadays are too delicate or too haughty, or they do not like the inconvenience."

Originally, the baby was taken to the wet nurse to be fed, but starting in the eleventh century, the nursemaid became a live-in domestic. She was usually better paid than other household servants. Often the mother alternated with the wet nurse so that she would have milk in the event that something happened to the nursemaid.

At first, only upper-class women used wet nurses, but as a middle class of shopkeepers emerged, especially in Italy, they too hired them, and different "authorities" in different countries had varying concepts of who would make the ideal nursemaid. A very popular English medical text from the first decade of the fourteenth century, *Rosa Medicinae*, by the contemporary doctor John of Gaddesden, but packed with advice from greats of the past—including Galen, Avicenna, and Averroes—recommended "a brunette with her first child, which should be a boy." Gaddesden also recommended that those suffering from what was called *phythisis*, tuberculosis, feed either from a wet nurse or directly from an animal's udder. This treatment for tuberculosis continued for centuries; ironically, it was not yet known that bovine tuberculosis can be transmitted from infected cows through their milk.

Bottle-feeding was not popular in medieval Europe, largely because it was believed that a baby nursed on animal milk would grow up to be less intelligent than one raised on human milk. But wet nurses also had their drawbacks. They were not always honest, and some were secretly heavy drinkers. Others sometimes supplemented their diminishing milk supply with goat's milk.

Many Europeans extended the belief that a baby takes on the characteristics of its wet nurse to a belief that a baby who suckles on animals will become wild and animal-like. For example, it was thought that a baby who suckled on a goat would become very sure-footed.

The medical historian Valerie Fildes suggests that the fact that the French were not particularly drawn to such beliefs explains why the practice of babies suckling on animals was more common in France. The philosopher and writer Michel de Montaigne observed in 1580 from his native Dordogne in southwest France, "It is common around here to see women in the village, when they cannot feed the children at their breast, call the goats to their rescue."

Samuel Pepys, the seventeenth-century English diarist, told a story in 1667 about Dr. Cayus, or Caius, a leading figure in one of Cambridge University's prominent colleges: the scholar was very old and lived entirely on women's breast milk. Dr. Thomas Moffet gave a similar account in his 1655 book *Health Improvement* and said that Cayus sucked the milk directly from a nursemaid's breasts. At first, apparently, his wet nurse was notably ill-tempered, and so was he, but then he switched to a better-natured wet nurse and became better-natured himself. It was not unusual in Europe for the sick and aged to be fed human breast milk.

Artificial feeding was rare. Many historians believe that this decision was often made with great pressure from the women's husbands. In fact, the entire breastfeeding issue was an early version of the modern abortion issue, with men trying to make the decisions about what women should do with their bodies.

★   ★   ★

TAILLEVENT MADE A sauce called *jance* from cow's milk. He did not specify how it was to be used, but jance was a popular sauce throughout medieval Europe, often served with poultry. There are many variations, and the only thing that they all have in common is the dominant taste of ginger. The word "jance" is thought to come from the word "ginger," which was one of the most valued products of the spice trade, largely controlled by the Portuguese. This is Taillevent's instructions for making milk jance:

COW'S MILK JANCE
Crush ginger, beat egg yolks, boil cow's milk, and mix them together.

Such sauces predate Taillevent. An anonymous 1290 manuscript, *Traité du XIIIe Siècle*, suggests:

The flesh of capons and hens is good roasted with a sauce of wine in the summer and in the winter, a sauce flavored with garlic and cinnamon and ginger, mixed with almond [milk] or ewe's milk.

This, in fact, was a sheep's-milk jance. Almost all the surviving recipe collections from the period offered jance recipes, but they were usually made with cow's milk. Ewe's milk was an unusual touch.

Given the risks of fresh milk, and the fact that it could not be used on the many holy days, substitutes had to be found. The most preferable was almond milk. That was why the anonymous recipe writer above gave jance makers a choice of using "almond or ewe's milk"; the almond milk was for the holy days.

Almond milk was made by grinding almonds to a fine powder, steeping the powder in boiled water, and then straining it. Recipes offering the option of using almond milk began appearing in the fifteenth and sixteenth centuries, but by the end of the sixteenth century, almond milk was already starting to disappear from European cooking. Why did it fall out of use? The risk of spoiled or bad milk certainly did not diminish. In fact, even as the British were dropping the use of almond milk at home, those living in Britain's tropical colonies, most notably Jamaica, were learning to substitute coconut milk for dairy milk, making it much the same way as they did almond milk.

One big factor in the move away from almond milk seems to have been the Protestant Reformation. The Protestants did not impose dietary restrictions for holy days, and as the restrictions faded away, so did the use of almond milk.

There was also another factor. Animal milk had become more easily available. During the Middle Ages, there were relatively few goats, sheep, and cows in Europe, and dairy farming was a small-scale family activity; it was not organized into a commercial enterprise until the eighteenth century. A farmer would have only two or three cows, or sometimes only one, and so did not have much milk to sell to others. Farmers were primarily interested in meat, and milk was a less important by-product. The purpose of keeping cows was that they bore calves. Cows were, to use the modern advertising label, grass-fed. But if you have only three cows, they do not need the extensive pastureland of a modern herd.

The famous cheese of Somerset, which dates back to medieval times, was made from clover-fed cows. This clover was grown in the field where the cows grazed. It was better to be a cow in pasture-rich England or Holland than in snowy Sweden or Norway, which is one of the reasons why England and Holland became the leading dairy countries. But even in England, and even in Somerset, cows withered away to a pitiful state in winter. Some died. No one had extra milk to share.

Furthermore, in medieval Europe and even considerably later, milk—which the English called "white meat"—was a seasonal food. As with all mammals, cows cannot lactate until they have offspring. That usually happened in the spring, and the cows could produce rich, plentiful milk through the spring and summer by eating fine warm-weather grasses. But in the winter, those grasses

Etching by Jean–Louis Demarne, 1752–1829. Milking a cow was one of many farm activities. (Author's collection)

died away. In addition, a cow spent its first two years, about a third of its life, as a heifer, meaning that for a long period it didn't produce any milk at all.

To get the maximum value from a lactation, a cow was generally milked for ten months of the year. But this was also a matter of timing because a cow carries its young for about nine months, and there is usually a two-month drying-off period between its pregnancies. The farmer has to make adjustments to keep a cow birthing in the spring to take advantage of the good grass.

Even cows whose lactation period stretched into winter did not produce much milk, so milk in abundance appeared only in the spring and summer. Fresh butter was available only in season, and only salted butter could be eaten the rest of the year. There was also something called "May butter," spring butter that was left to cure unsalted; it was considered to be highly medicinal.

The seasonal nature of milk gave it a special feeling—it was one of the glories of spring and summer. Unfortunately, though, it also meant that milk was produced in warm months, when it was most likely to spoil.

Often, if a farmer lived near a city, and most did because it was an obvious advantage for a fast-spoiling product, he would take his cow into town to sell its milk. This was particularly common in London. The farmer would wander the streets, calling out, and women, either housewives or servants, would come out with buckets or other receptacles and he would milk the warm, foaming liquid directly into their containers. This had a number of advantages. The milk could not be fresher, and the customer could inspect the cow, to make sure it was healthy and well cared for.

As London expanded, it established parks such as Hampstead Heath, originally a separate village, St. James's Park, and Lincoln's Inn Fields and allowed cows to graze in them. St. James's Park had a reputation for high-quality milk, and affluent Londoners sent their staff there to collect it. But the move away from the door-to-door selling of fresh milk marked the beginning of what became a dangerous decline in the quality of urban milk. Women and milk-maids would milk the cows in the parks and carry two pailfuls hanging from a bar across their shoulders through the town. Soot, sticks, and other urban detritus might fall into these uncovered pails, and the washing of the pails between milkings was not common practice. Similarly, after dairies were established near towns and cities, milk was transported in uncovered pails or exposed panniers hung on donkeys. Some historians have noted a decline in the status of milk among urban people in the sixteenth century. Like the ancient Romans, they began to think of it as food for the poor, and it may have been the lack of hygiene surrounding milk—arriving in open vessels with twigs, bugs, and dirt floating on top—that was the reason for this distaste.

Markham's 1615 book for English housewives was the first English book that dealt with proper dairy management. He did emphasize cleanliness. But as often happens when men lectured women, the subject of hygiene got jumbled up with the urge to judge women:

> Touching the well ordering of milk after it has come home to the dairy, the main point belonging thereunto is the housewife's cleanliness in the sweet and neat keeping of the dairy house; where not the least mot of any filth may by any means appear but all things either to the eye or nose so void of sourness or sluttishness, that a prince's bed chamber may not exceed it.

On the other hand, water was usually no cleaner than milk. The safest drink was beer. This may be why a number of northern cultures mixed milk and beer. It is an old idea in many northern countries. Milk and beer are the central ingredients in some early possets, and in several countries, including Holland and Scotland, oat porridge was commonly mixed with either ale or milk and often both. In what is today southern Sweden but used to be Denmark—the provinces of Skania, Halland, and Blekinge—beer and milk were commonly consumed at breakfast and most other meals. The beer was a type called small beer, meaning that it had a very low alcohol content. A mug of equal parts of small beer and milk, along with a slice or two of dark bread, was often a poor man's meal. Beer and milk served with herring was also a common breakfast. The milk was sometimes fresh whole milk, sometimes skimmed milk, or sometimes even sour milk.

The idea of milk mixed with ale has persisted. In 1875 John Henry Johnson of Lincoln's Inn Fields, a place with a milk history, requested a patent for a light alcohol beer made from whey, lactose, and hops. He never made the beer, but others did, or else made similar concoctions that they called milk stout. Eventually the only milk used in milk stout was lactose, and the British government, reasoning that this did not offer the health benefits people expected from milk, declared in 1946 that the word "milk" could no longer be used in the name.

7

THE CHEESE HEADS

Fresh and young soft cheeses were popular with ancient Greeks and Romans and remained so in Europe. Originally they were almost always made of goat's or sheep's milk, but toward the end of the Middle Ages, there was a tendency to use cow's milk. At first, such cheeses were made of pressed curds, with perhaps some local herbs added, and briefly aged. But in time, they became increasingly elaborate. This one is from Gervase Markham's 1615 book:

TO MAKE A FRESH CHEESE

To make an excellent fresh cheese take a pottle [half gallon] of milk as it comes from the cow, and a pint of cream: then take a spoonful of rennet or earning and put it unto it, and let it stand two hours, then stir it up, put it into a fine cloth, and let the whey drain from it: then put it into a bowl and take the yolk of an egg, a spoonful of rose water, and beat them all together with a very little salt, with sugar and nutmegs, and when all these are brayed together and searced [completely blended], mix it with the curd, and then put it into a cheese vat with a very fine cloth.

Cream cheese was also popular. Today we often eat cream cheese, but it is so industrialized that we have lost sight of what it is, or what it was. It dates back to at least the fifteenth century, and was very popular in numerous countries, including Scotland, England, and France. This is Eliza Smith's 1727 cream cheese recipe from her book *The Compleat Housewive*, which was printed in eighteen editions, many of which appeared after her death. Smith's book was the first cookbook printed in colonial America. Note that this recipe is called Summer Cream Cheese. That's because summer was the time of year when you could get three pints of milk straight from the cow.

TO MAKE SUMMER CREAM CHEESE

Take three pints of milk just from the cow, and five pints of good sweet cream, which you must boil free from smoke, then put it into your milk, cool it till it is blood warm, and then put in a spoonful of rennet, when it is well come [set] take a large strainer, lay it upon a great cheese fat, then put the curd in gently upon the strainer, and when all the curd is in, lay on the cheese board, and a weight of two pounds: Let it so drain three hours, till the whey be well drained from it, then lay a cheese cloth in your lesser cheese fat, and put in the curd, laying the cloth smooth over as before, the board on the top of that, and a four pound weight on it, turn it every two hours into dry cloths before night and be careful not to break it next morning. Salt it in the fat until the next day, then put it into a wet cloth which you must shift every day till it is ripe.

The switch from making sheep's-milk or goat's-milk cheese to making cow's-milk cheese was in part because as cheesemakers became more commercial in outlook, they wanted to produce a cheese that could travel well and withstand some abuse. That meant hard, aged cow's-milk cheese. Soldiers needed this kind of cheese as well. Detailed accounts remain of the provisions used by some of the soldiers in the first English Civil War, 1642–46. The two hundred men on the parliamentary side garrisoned in Wiltshire were provided with 5,300 pounds of cheese and 400 pounds of butter. The foot-soldier contingent received 16 pounds of cheese per day, along with 8 1/2 pounds of butter, 13 pounds of bread, and 40 pints of beer. Beer was deemed critical for maintaining stamina.

Before long, one cheese began dominating the international cheese trade: Parmesan, which endures to this day. One of the earliest records of this cheese is from a ledger showing that some Parmesan was purchased by the commune of Florence in 1344. However, the cheese is thought to have begun much earlier than that.

Parmesan became renowned after 1357, when Giovanni Boccaccio, a Florentine writer, in his celebrated book of stories, *Decameron*, wrote a fantasy about a mountain of grated Parmesan where "dwell folk that do ought else but make macaroni and ravioli." The writer set this fanciful mountain far from Italy "in the land of the Basques." But no doubt he would not have made such a reference unless the cheese was already well known in southern Europe.

Parmesan cheese was also called Parmigiano-Reggiano, because it was, and still is, made in the rich green pastureland between Parma and Reggio. Watered by the Po River, it is Italy's great cow region, Emilia Romagna. If you see a pasta dish made with butter, it probably comes from Emilia Romagna, the one region in Italy that uses butter rather than olive oil.

To make Parmesan cheese, the cheesemaker left late-afternoon milk overnight. Then, in the morning, he skimmed the milk and made butter with the cream, while mixing the leftover skim milk with the fresh milk brought in from the morning milking. Then rennet and some whey from the cheesemaking done the day before were added. The mixture was heated to a low temperature for forty minutes, at which point it was curdled and the whey removed. The excess whey was fed to local pigs, which produced Italy's most famous ham, prosciutto di Parma. To bear that name even today, the ham must come from a pig that is fed Parmesan whey.

Parmesan was made into very large wheels and aged for two or three years. Its fame endured. Two centuries after Boccaccio wrote of a mountain made of grated Parmesan, the Renaissance food writer Platina singled the cheese out for its quality. Samuel Pepys claimed that he had saved his supply from the London fire by burying it in his backyard. Thomas Jefferson had Parmesan cheese shipped to him in Virginia.

By Jefferson's time, it was not unusual to import Parmesan from Italy. William Verrall, an eighteenth-century chef in Lewes, Sussex, England, in his 1759 cookbook gave this recipe for what in time would be known as macaroni and cheese. Verrall had studied under a French chef and apparently learned about Parmesan from the French without knowing much about Italian food. He explained in a recipe previous to this one, for "macaroons in cream"—a devastatingly rich dish of sugar, egg yolks, cream, and butter—that these macaroons are not "the sweet biscuit sort but a foreign paste, the same as vermicelli, but made very large in comparison to that."

MACAROONS WITH PARMESAN CHEESE

For this too you must boil them in water first, with a little salt, pour to them a ladle of cullis [broth], a morsel of green onion and parsley minced fine, pepper salt and nutmeg: stew all a few minutes, and pour into a dish with a rim as before, squeeze a lemon or orange, and cover it over pretty thick with Parmesan cheese grated very fine, bake it to a fine colour about as long a time as the last [quarter of an hour], and serve it up hot.

The French serve to their tables a great many dishes with this sort of cheese, and in the same manner, only sometimes with a savoury white sauce, such as scallops, oysters, and many of the things you have among these entremets.

Another internationally famous cow's-milk cheese is cheddar, originally from the town of Cheddar in pasture-rich Somerset, England. Unlike Parmesan, however, the place of origin of cheddar cheese has never been legally guaranteed. It is not nearly as old as Parmesan, though it dates back to at least the sixteenth century and probably earlier. The word "cheddar" originally referred to the town, but soon came to mean the cheesemaking process. "Cheddaring"

"Cutting up curd," The Illustrated London News,
November 4, 1876.

means slicing partially strained curd, stacking it, and then turning
and restacking it every ten minutes during the pressing process. This
makes cheddar an unusually smooth cheese.

Like Parmesan, cheddar cheeses are very large cheeses, though
few are as large as the nine-foot-diameter, 1,250-pound cheddar made
for Queen Victoria's wedding. In Somerset, cheddars were called
"corporation" cheeses because it would take the milk of an entire
parish of farms to make one cheddar. The Queen's cheddar took two
parishes. The cheese also needed to age between one and two years,
so making these large cheeses was a real investment of time and
money. Smaller cheddar cheeses, known as truckles, were sold locally,
but the big cheeses were designed to travel. The idea of cheddar was
also exported and the cheese was made in most British colonies.
There are Irish, Canadian, American, and Australian cheddars, but
rarely do any of these have the distinctive "nutty" flavor of a Somerset
cheddar. Few seem to notice.

Eliza Smith's book included a recipe for making cheddar, contrib-
uting to the idea that it could be made anywhere, although clearly
you would have to be on a farm to do so. Cheesemakers may have

started thinking in commercial terms, but cheesemaking was still usually a family activity, and cookbooks of the seventeenth and even eighteenth century often included various cheese recipes. Smith's book, which was intended for housewives, gives recipes for making rennet, butter, and several kinds of cheese.

TO MAKE A CHEDDAR CHEESE

Take the new milk of twelve cows in the morning, and the evening cream of twelve cows, putting to it three spoonfuls of rennet: when it is come, break it and whey it [drain it]; that being done break it again, that being done work into the curd three pounds of fresh butter, put it in your press, turn it very often for an hour or more, and change the cloths, washing them every time you change them, you may put wet cloths at first to them, but toward the last put two or three dry cloths, let it lie thirty or forty hours in the press, according to the thickness of the cheese, then take it out, wash it in whey, and lay it in a dry cloth until it is dry, then lay it on your shelf and turn it often.

THE GREAT DAIRY consumers lived in northern Europe and in the Alps, especially Switzerland. Despite the fall of Rome, southern Europe had never lost its sense of superiority over its neighbors and still considered them to be milk-swilling barbarians. Southerners did drink milk and certainly ate cheese, but never on the same scale as the northerners. The Dutch in particular were singled out as a crude and comic people endlessly engorged on milk, butter, and cheese. Even the Flemish laughed at them, calling them *kaaskoppen*, "cheese heads." Northerners, too, especially the English, belittled the Dutch for their dairy habits. One English pamphlet said, "A Dutchman is a lusty, fat, two legged cheese-worm."

More than mere cheese heads, the medieval Dutch were also great porridge heads, enthusiastically consuming quantities of a very pleasant variation of what the French and English were calling

"Milking Time" engraving by John Godfrey from the Art Journal of 1856, from a 1649 painting by Paulus Potter.

frumentum, a Latin word for "wheat." Typically, the Dutch ate porridge at least once a day. It could be eaten at breakfast, lunch, or dinner, and as an appetizer, main course, or dessert.

A 1560 book, *Nyuwen Coock Boek*, gave this recipe for Dutch porridge, which seems well worth copying:

> Boil hulled grains of wheat in water and drain them. Next, boil plain milk with egg yolks, sugar, and saffron. Add grains of wheat and let them boil for a while.

Boiling with egg yolks was a popular Dutch technique for curdling milk. Porridge was becoming less a cereal and more a sweetened custard or pudding.

In addition to milk, cheese, and porridge, the Dutch ate huge quantities of butter. The upper classes would pride themselves on setting their tables with several types of butter, perhaps one from Delft, one from Gouda, and others. The Dutch preferred cooking with butter to cooking with lard as other northerners did. They also enjoyed whey or buttermilk with breakfast. Even in the poorhouses, breakfast was buttermilk and bread.

And what was a Dutch meal without butter? Butter was used wherever possible. Even *hutsepot*, a traditional meat stew, used butter. This is a 1652 recipe offered by a doctor, Jan van Beverwijck from Dordecht:

> Take some mutton or beef, wash it clean and chop it fine. Add there to some green stuff or parsnips or some stuffed prunes and the juice of lemons or oranges or citron [resembles a lemon but much larger, usually only the peel is eaten], or a pint of strong, clear vinegar, mix these together, set the pot on a slow fire; add some ginger and melted butter and you shall have prepared a fine *hutsepot*.

You would probably need to cook the *hutsepot* for about four hours.

The Dutch navy, which in the sixteenth century was becoming a formidable force, issued to each sailor a weekly ration of half a pound of cheese, half a pound of butter, and a five-pound loaf of bread. The historian Simon Schama calculated that a Dutch ship with a crew of one hundred in 1636 would need among their provisions 450 pounds of cheese and one and a quarter tons of butter. An ample supply of cheese and butter were the right of every Dutchman.

The French told "Flemish jokes" as they do "frites jokes" today, but rather than ridiculing them for their fried potatoes, they laughed at them and the Dutch for the extent of their butter consumption. Foreign mockery aside, Dutch health authorities constantly complained that the Dutch ate unhealthful quantities of butter. Particularly disturbing to many was the Dutch habit of eating butter and cheese together. In the mid-seventeenth century, the painter Paulus Potter wrote:

> *Cheese with butter is an evil*
> *Wished upon us by the Devil*

But mostly the Dutch did believe that dairy food was an essential part of a good diet. The seventeenth-century doctor Heijman Jacobs advised consuming "Sweet [fresh] milk, fresh bread, good mutton and beef, fresh butter and cheese." Artists from the celebrated Dutch school of still-life painting often included cheeses in their compositions.

The Dutch made numerous types of cheese—some with sheep's milk, but more with cow's. They even made a cheese from colostrum, *caseum nymolken*. They made a new spring cheese, a cheese with cumin, and cottage cheese, which is curds with a small amount of whey left in. They also made imitations of English cheese, French Brie, and Parmesan. In the sixteenth century they started making Italian-style ricotta.

The Dutch had an effective distribution system for cheese, with numerous urban centers featuring cheese markets. It was a small country, and most people lived on or near a farm or near a cheese market. Gouda cheese (pronounced *how*-da in Dutch, with a guttural *h* like the Spanish letter *jota*, which is where it comes from) was named after the town in whose cheese market it was sold. The earliest record of this cheese was in 1184, and as is usually true of first records, the cheese had probably already been made for some time before this date. Dutch cheesemaking itself certainly goes back long before this. When the Romans first came to the Dutch lands in 57 B.C.E., according to Julius Caesar, they found a cheese-eating people. The Dutch word for cheese, *kaas*, comes from the Latin *caseus*, which has led some historians to conclude that the locals learned to make hard cheese from the Romans; the Roman soldiers had hard cheese rations for marching.

Gouda is an aged cow's-milk cheese, and like many cheeses, it was always made by women at the farmhouse, despite requiring the hard manual labor of stirring and cutting curd in huge vats. The milk was brought in still warm from the two daily milkings and was curdled not only with calves' rennet, but also with lactic acid bacteria cultured by the cheesemaker.

Though most Gouda today is made in highly industrialized factories, there are a few farms that make it much the same way it was made in the twelfth century, with raw milk fresh from the cow. This cheese is labeled *boerenkaas*, farm cheese, and it is collected from the farms by wholesalers.

The Dutch began exporting cheese in the thirteenth century. By the sixteenth century, they had such a surplus that they were exporting it to England, France, Germany, Scandinavia, Spain, and Portugal.

In the thirteenth and fourteenth centuries, the Dutch became skilled at reclaiming land from the sea by building dykes and creating polders, drained patches of reclaimed seabed. This led to dramatic improvements in cattle breeding and land maintenance. A great portent for the future was the fact that in the northwest, just below the North Sea in the province of Friesland, the only part of Holland

The cheese market at Hoorn, a town in North Holland, from an 1880 issue of La Tour du Monde, *a popular French travel weekly. The artist, Ferdinandus, was a popular illustrator for the magazine.*

to have its own language, farmers were having tremendous success crossbreeding livestock to develop cows that produced more milk.

Between the mid-sixteenth and mid-seventeenth centuries, the value of a Dutch cow quadrupled. The Dutch were starting to understand what best to feed cattle, as well as how best to cultivate pastureland. This led to an enormous increase in milk production in Friesland, Flanders, and Holland in the sixteenth and seventeenth centuries. The Dutch cows were producing more than twice the yield of cows in neighboring countries, and milk was more plentiful in the Netherlands than in most of Europe.

Though at first unnoticed, a huge shift was taking place in the European perception of the Dutch. Their country, which had broken off from Spanish rule in the 1590s, was rapidly shifting from a former lowly possession of the Holy Roman Empire to an independent republic displaying a genius in art, science, and engineering. Seemingly overnight, the Netherlands was becoming a global trading empire and the leading maritime and economic power of the world. Suddenly, the cheese heads were considered brilliant.

All over Europe there were discussions and writings about what made the Dutch such geniuses. And those having these discussions often freely admitted that they had once thought of the Dutch as idiots who just drank milk and ate cheese. The Europeans had also started recognizing that there was genius in the Dutch dairy farms— in their better pastures, better cows, and ability to farm below-sea-level land. Dutch dairying, too, was now considered brilliant.

AN EVEN BIGGER shift in the dairy world was also taking place. The Dutch, English, French, Spanish and Portuguese, milk drinkers all, driven by unabashed greed and a sense of adventure, were undertaking long-distance sea voyages to faraway lands—two vast continents—where humans did not milk animals.

8

TO MAKE PUDDING

THERE WERE NO cows or cattle of any kind in America until the Europeans brought them. There were a number of breeds of goats and sheep, but no one milked them. No one tried to domesticate or milk the bison of North America either. The camel-like llama of South America produces excellent milk and is easily farmed, but they were never milked, even though they lived near the Incas, a great and advanced civilization. Llamas were domesticated, but used only for carrying, while a smaller breed, alpacas, was used for its wool. But from the Arctic Brooks Range of Alaska to the Antarctic tip of Patagonia, there was no dairying. Most people were lactose-intolerant.

In the Americas, humans, like all other mammals, provided their own milk to their young. In fact, in some cases when milk was needed for the survival of a nonhuman animal, women provided it. This has been documented among some Canadian tribes, and the Puma of the Sioux nation, and some Amazonian tribes. But it is not unique to the Americas. It had sometimes occurred in Europe and Asia, including ancient Rome as well. There are also records of women in the highlands of New Guinea breastfeeding piglets, pre-European Hawaiians breastfeeding puppies, and Guyanese women breastfeeding deer in South America. Women breastfed animals whose survival was important to them, and also to stimulate lactation and relieve engorged breasts.

Soon after Europeans arrived in the Americas, they brought over cows. They were not going to live without milk, and there was no

real debate about what kind of milking animal to bring, because if you are going to bring milk producers over on long ocean voyages, it is logical to bring the animal that produced the most milk per head.

THE DIFFERENT EUROPEAN nationalities that came to the Americas had two things in common: They all ate beef and dairy products, and they all wanted to make life in the new land resemble life in the old land as closely as possible. This is in large part why they tended to view the people and cultures living in the Americas as simply being in the way. They did learn to eat some local products, such as corn and turkey, but mostly transported European food over as quickly as possible.

On Christopher Columbus's second voyage to the Americas, he brought cattle to Santo Domingo, which is now the Dominican Republic. They did not resemble European breeds and are thought to have descended from Indian stock. Cows from India had probably been brought by the Arabs to Spain, where they were interbred with European stock. In 1525 the Spanish brought cattle to the Mexican Caribbean port of Vera Cruz. The animals eventually spread throughout Mexico and are thought to be the origin of the Texas Longhorn. The Spanish also brought cattle from the Canary Islands to South America. Some sheep and goats were brought over, too.

Cows were not an overnight success. They were attacked by mosquitoes, killed by warring tribesmen, and slaughtered by starving farmers for their beef. Dairying struggled for more than a century before becoming central to the South American diet. Eventually, though, missionaries on the west coast of the continent were able to live off the milk and vegetables they produced, and they taught the native women to make cheese and butter.

The first cattle in North America were brought to Florida, Georgia, and the Carolinas, also by the Spanish. The British brought their first cows to a new colony on Roanoke Island in 1580, but the colony, cows and all, disappeared. When the British founded their second Virginia colony, Jamestown, in 1606, three ships brought over 140 colonists, but no cows. Ships supplied the colonists with cheese and butter for a time, despite a British law prohibiting the export of butter, and then, just as the colony was dying out in 1610, Lord Delaware sailed up the Chesapeake and saved it with more men, provisions, and milking cows.

Lord Delaware believed that the Jamestown colony had to develop dairy farms if it were to survive. When he returned to England in 1611, he wrote a report to the Virginia Company, explaining that "the Cattell [cattle] already there are much increased, and thrive exceedingly with the pasture of the country, the kine [cows] all this past winter, though the ground was covered most with snow, and the season sharp, lived without other feeding than the grass they found, with which they prospered well, and many of them readie to fall with Calve."

He went on to write, "Milke being a great nourishment and refreshing to our people serving as well for Physicke [health] as for food, so that it is in no way to be doubted, but when it shall please God that Sir Thomas Dale and Sir Thomas Gates shall arrive in Virginia with their extraordinary supply of one hundred Kine."

Thomas Dale, the new deputy governor of the colony of Virginia, had modern ideas about dairy farming. While today the progressive approach might be to grass-feed cows, in the early seventeenth century, it was unusually advanced to believe in using supplementary feed. Dale did believe that cows should be specially cared for in high-quality pastures, but he also thought they should have supplemental food and shelters. Dale built the first cow barn in the future United States. He also arranged for hay to be grown, harvested, and stored for winter food. He ruled by martial law and made the wanton killing of a cow a capital offense.

But when Dale left the colony in 1616, leaving behind more than two hundred head of cattle, most of his ideas were abandoned. His cow barn was used for storage, and the cows no longer had either shelter or winter feed. Captain John Smith, a key figure in the colonization of both Virginia and New England, was better known for his exploits with women than for farming. He found the idea of feeding cows ridiculous and believed they should be left to forage.

However, even without Dale, the Virginia Company did start to recognize the importance of dairy farming. A Virginia cow was priced at more than twice the price of a cow in England, and the company adopted a policy of shipping twenty heifers for every one hundred new settlers. By 1629 the colony had more than two thousand people and possibly as many as five thousand head of cattle. When Virginia became a democracy in 1619, they adopted laws against killing cows, one of Dale's few surviving ideas. In 1673, during an unusually harsh winter, thousands of cows without shelter or feed died. Nonetheless, practices in the American dairy industry, though frequently argued about, did not change, and most cows had no shelter or supplemental feed until the nineteenth century. And the debate about feeding cows has continued.

WHEN THE NEXT group of English colonists settled in what the English called Northern Virginia, what John Smith labeled New England, and what local people called Massachusetts in 1620, it seems logical to assume that they would have learned something from the failures and successes of the Jamestown colony. But instead, they had to starve before learning the exact same things that the Virginia colonists had learned.

The people who landed in Plymouth in 1620 were different from the ones who had settled in Virginia. They were religious extremists looking for a place to experiment with their ideas of theocracy. They cared little for details such as food supplies and so not only brought very little food with them, but also lacked skilled farmers and fishermen, nor did they bring the necessary farming and fishing equipment.

Butter was one of the few foods with which they were amply supplied, at least initially—a hint of the standing of butter in early-seventeenth-century England. But apparently, William Bradford, the governor of the new colony, did not believe in the importance of butter, and when they found themselves short the necessary funds to leave England, he raised the difference by selling most of it off, calling it "a commodity they might best spare."

So the colonists paid their port duties and sailed off with scant food supplies and almost no knowledge of farming or fishing. One pilgrim wrote, "beseeching the Lord to give a blessing to our endeavore, and keep all our hearts in the bonds of peace and love, we take leave and rest."

They apparently did have at least some dairy provisions, however, because on November 26, two months and ten days after landing, sixteen men, well armed but with little food except for a provision of Dutch cheese, set out to explore Cape Cod.

Even with some of the local Native Americans helping them, the settlers starved the first winter. Half of their 101 or 102 members died.

With their remaining provisions, that March they entertained an Indian leader, Samoset, with biscuits, butter, cheese, and wild ducks. Samoset reportedly enjoyed the dairy products, which he had sampled before from English fishing ships. Finally, in March 1624, an English ship, the *Charity*, arrived with one bull and three heifers, and the New England dairy industry began. Subsequent ships brought more and more cattle of various breeds—black, red, and red-and-white. The black cows are thought to have been Kerry, a popular Irish breed descended from early Celtic cattle.

In 1639 the first settlers arrived in Boston with thirty cows. More were brought in soon thereafter, and Boston developed a dairying culture. By 1650, butter and cheese were being exported from the New England colonies.

THE EARLY DUTCH settlers were different. When they established their North American colony in New York in 1624, their first shipload of settlers brought along cows, as well as sheep and pigs. And in 1625, yet more cows, sheep, horses, and pigs arrived. Ships with names such as *Cow* and *Sheep* and *Horse* were specially designed to transport livestock.

After arriving, the cows were quickly ferried from what is now Governors Island to Manhattan, which had good pastureland. But despite this, the colony's religious leader, Jonas Michaelius, wrote back to Amsterdam that they did not have enough milk or butter and needed more cows. More black-and-white cows from Friesland were sent. In 1639 anyone who was willing to settle in what was now called New Netherlands was given free passage, a house, a barn, farming equipment, four horses, four cows, sheep, and pigs. After six years they were obligated to give back four cows, but by then they were expected to have bred more than that.

De Verstandige Kock, translated in 1989 by the Dutch American food writer Peter G. Rose as *The Sensible Cook*, was first published in Holland in 1667 and became the leading cookbook of the New Netherlands colony. One recipe uses cream and eggs to make a rich porridge:

> Take 12 Egg yolks, a pint of cream, pour the Eggs through a sieve and mix well with the Cream, add to it Rosewater and Sugar appropriately, place it on the fire, stir it gently until it thickens, but do not boil or it will separate.

A number of other equally egg-rich custards were also included in the cookbook. This was the lemon custard:

Take the juice from the Lemons and the yolks from 8 Eggs, but add only the white of four, grate a White-bread of half a stuyver [2.5 cents' worth, though it is difficult to say how much bread half a stuyver might have bought—a small loaf seems a good guess], then a pint of Sweet Milk and Sugar proportionately, neither too vigorously or too slow you should let it boil.

Yet another recipe used buttermilk to make an applesauce, called apple-a-milk:

Take Aeghten [a variety of sour apple] Apples peeled and with the cores well removed, place them in a pot with some Butter and Rosewater. Let it cook until it is fine like porridge, mash it steadily with the spoon, then add a little Wheat-flour, add a proportionate amount of Buttermilk. Let it cook together until it is like sweet cream, then add some Sugar and White-bread.

BUTTERMILK WAS WELL liked in the North American colonies, although in the South it was mostly given to slaves. Cottage cheese was also made. Originally it was a product for home consumption, but it grew in popularity and eventually became a mainstay of the American diet, though "popular" may not be the right word. It earned a reputation as a diet food, and many people ate it without really liking it. In 1968, when Richard Nixon was running for president, his campaign managers were looking for opportunities to make him appear more human and endearing. For Nixon was one of the least cuddly of politicians, stiff and often ill-tempered, with a tendency to go off in strange directions when least expected. While campaigning in Oregon, Nixon went on statewide television and took questions from ordinary people. One question was "How do you keep your weight down?" Partly by eating cottage cheese, he said. But he confessed to hating it:

> I eat cottage cheese till it runs out of my ears. But I've learned a way to eat it that makes it not too bad. I put ketchup on it. At least that way it doesn't taste like cottage cheese. I'll tell you where I learned that—from my grandmother. My grandmother lived to be ninety-one years of age, and she always mixed ketchup with her cottage cheese.

But of course cottage cheese doesn't have to be low in fat. The Waldorf-Astoria's Swiss-born chef Oscar Tschirky, the most famous chef in New York in the late nineteenth century, gave the following advice on how to make cottage cheese. It has little to do with simple, partially drained curds, which is the way cottage cheese is normally made:

> A richer way is to put equal parts of buttermilk and thick milk into a kettle together over the fire, heat it until nearly ready to boil, pour into a linen bag and let it drain until the next day. Then remove, salt, and put in a little cream, or butter, according to whether it is thick or not and make up into balls the size of an orange.

SPOILAGE WAS A constant problem. In America in the seventeenth and eighteenth centuries, buckets of milk or cream were lowered into wells to keep them cool, and butter and cheese were kept in cold cellars.

Butter turning rancid was an issue. Catherine Beecher's 1869 book, *American Woman's Home*, which she mostly wrote herself, though it also bears the name of her more famous sister, the abolitionist Harriet Beecher Stowe, describes how an entire meal can be ruined by bad butter:

A matter of despair as regards bad butter is, that at the tables where it is used, it stands sentinel at the door to bar your way to every other food. You turn from your dreadful half-slice of bread, which fills your mouth with bitterness, to your beef-steak, which proves virulent with the same poison; you think to take refuge in vegetable diet, and find the butter in the string beans, and polluting the innocence of early peas; it is in the corn, in the succotash, in the squash; the beets swim in it, and the onions have it poured over them. Hungry and miserable, you think to solace yourself at the dessert; but the pastry is cursed, the cake is acrid with the same plague. You are ready to howl with despair and your misery is great upon you—especially if this is a table where you have taken board for three months with your delicate wife and four small children. Your case is dreadful, and it is hopeless, because long use and habit have rendered your host incapable of discovering what is the matter. "Don't like the butter, sir? I assure you I paid an extra price for it, and it's the very best in the market. I looked over as many as a hundred tubs, and picked out this one." You are dumb, but not less despairing.

Many cookbooks offered dubious solutions. *Mrs. Hill's Southern Practical Cookery and Receipt Book*, an 1867 book by Annabella P. Hill, recommended:

To recover rancid butter or lard—Use Darby's prophylactic fluid by the directions that accompany the bottles; cream it thoroughly, then put the butter in a clean vessel.

Darby's prophylactic fluid had been invented by John Darby, and according to him, this panacea contained hypochlorite of potassa (potassium hydroxide) and a cocktail of sodas and salts. It was widely used in the South.

Many nationalities weighed in on the butter issue. Elena Molok-hovets's *A Gift to Young Housewives*, first published in 1861, was the leading cookbook of the prerevolutionary Russian aristocracy. She offered numerous recipes for "rectifying" spoiled butter:

Thoroughly wash the spoiled butter in several changes of water. Add salt and juice from carrots that have been grated and squeezed through a cloth. Mix until completely incorporated.

Carrot juice gives the butter a very delicate, pleasant taste but it is better to add it just before serving, because this butter will only keep for a few days.

It is significant that the first attempt to build a refrigerator in the United States was undertaken for keeping butter (as opposed to what happened in Australia, where in 1853 James Harrison, a Scot who had moved there, developed the first really fully functional refrigerator and used it for chilling beer). In 1803, Thomas Moore of Maryland built a metal box for butter surrounded by ice that was held in an outer cedar box insulated with fur. He carried butter in his box, which he called a refrigerator, the first use of the word, the twenty miles from his Maryland farm to the food market in Georgetown. At the market, his hard, fresh butter was a sensation and shoppers gladly paid a higher price for it. But Moore failed in his ultimate goal of creating a refrigerator industry. People wanted to buy the butter, but not the refrigerator (though Thomas Jefferson did buy one).

Spoilage was not the only problem. In 1863, Isabella Beeton, an Englishwoman married at age nineteen and dead from the birth of her fourth child at age twenty-nine, published *Mrs. Beeton's Book of Household Management*, one of the most widely read food books in the history of the English language. She pointed out that butter can have a very "agreeable flavor." But it all depended on what the cows are being fed, and some cows produced "bad butter."

Butter and cheese were made on the farm by women. It was hard physical labor that took long hours. Lisa Smith described the process in the eighteenth century:

As soon as you have milked, strain your milk into a pot, and stir it often for half an hour, then put it in your pans or trays; when it is creamed. Skim it exceedingly clean from the milk, and put your cream into an earthen pot, if you do not churn immediately for butter, shift your cream once in twelve hours into another clean scalded pot, and if you had any milk at the bottom of the pot, put it away; when you have churned with your butter in three or four waters and then salt it to your taste, and beat it well, but do not wash it after it is salted, let it stand in a wedge, if it be to pot, till the next morning, and beat it again, and make your layers the thickness of three fingers, and then throw a little salt on it, and so do till your pot is full.

It is the "churned," "beat it well," and "beat it again" aspect of this recipe that is exhausting.

"Churning butter," The Illustrated London News, *November 4, 1876.*

A device was invented to use the motion of rocking a cradle to help churn butter. A balance bar was hooked up to a cradle on one end and a butter churner on the other. As the mother gently rocked the cradle, perhaps with one foot so her hands were free for another task, the butter was being churned.

FROM JAMESTOWN AND Plymouth until after the Civil War, American food was basically English. This was why they rarely made any cheese other than cheddar, and if they did, it was probably Wiltshire. In the colonial period, Americans followed the popular English cookbooks of the time—by Gervase Markham, Hannah Glasse, and especially Eliza Smith, whose book was published in Williamsburg, Virginia, in 1742, a decade after she had died. There were some additions to the American edition of Smith's book, including a remedy for rattlesnake bites and a "Cure for Poison." These remedies were given to the book's publisher by a man named Caesar, who was a black slave. He was granted his freedom and £100 a year for the rest of his life in exchange for the remedies.

"The first American cookbook," so called because it came out in 1796, after the War of Independence, was published in Hartford, Connecticut, and authored by Amelia Simmons, presumably an American, though almost nothing is known about her, including her place of birth. She did call herself "an American Orphan," but most of her recipes are clearly English, despite the claim on the title page that they are "adapted to this country." However, her syllabub recipe, which calls for directly milking from the cow, uses cider, which does seem American:

TO MAKE A FINE SYLLABUB FROM THE COW

Sweeten a quart of Cyder with double refined sugar, grate nutmeg into it, then milk your cow into your liquour, when you have thus added what quantity of milk you think proper, pour half a pint or more, in proportion to the quantity of Syllabub you make, of the sweetest cream you can get all over it.

Simmons offered four recipes for "creams," which was significant in a recipe book that was only fifty pages long. Creams, which were more like what we would call today a mousse, were an eighteenth-century fashion. Eliza Smith gave fifteen cream recipes in her book, and Hannah Glasse sixteen. Simmons chose to include "a fine cream" and a lemon, raspberry, and "whipt" cream. This is her fine cream:

Take a pint of cream, sweeten it to your palate, grate a little nutmeg, put in a spoonful of orange flower water and rose water and two spoonfuls of wine; beat up four eggs and two whites, stir it all together one way over the fire till it is thick, have cups ready and pour it in.

This is a fairly simple recipe compared to some of the English creams. Eliza Smith had the same idea as Simmons, which she called "blanched cream"; it was probably the origin of Simmons's recipe:

Take a quart of the thickest sweet cream you can get, season with fine sugar and orange flower water; then boil it, then beat the whites of twenty eggs, with a little cold cream, take out the treddles [meaning dung or perhaps dirt?], and when the cream is on the fire and boils, pour in your eggs; stirring it very well until it comes to a thick curd; then take it up and pass it through a hair sieve [circular wooden frames with woven horsehair screens were a common kitchen tool for centuries]; then beat it very well with a spoon till it is cold and put it in dishes for use.

An unusual and intriguing variation of a cream was offered by Elizabeth Raffald, a domestic from Yorkshire, in her 1769 book, *The Experienced English Housekeeper*. She called it La Pompadour cream:

Beat the whites of five eggs to a strong froth, put them into a tossing pan, with two spoonfuls of orange flower water [and] two ounces of sugar, stir it gently for three or four minutes, then pour it into your dish, and pour good melted butter over it, and send it in hot.—It is a pretty corner dish for a second course at dinner.

Simmons also had a separate section on puddings, which was not unusual for seventeenth-, eighteenth-, or even nineteenth-century cookbooks. She gave twenty-nine pudding recipes. Eliza Smith included fifty-six pudding recipes. Robert May, who cooked for Royalists and Catholics in England's turbulent seventeenth century, published *The Accomplisht Cook* in 1660 and again in 1680. Considered the definitive work on the food of Charles II's court and the restoration of monarchy, it includes fifty pudding recipes.

While a pudding eventually became a milk-based dessert, originally it was neither. The early puddings were often made with meat or meat sausages. Suet, butter, veal, mutton, and dried fruit were common ingredients, stuffed together into a sheep's stomach or intestines. The English still call blood sausage "black pudding." In Robert May's 1660 "pudding in wine and guts" recipe, "guts" means a sausage casing:

Take the crumbs of two manchets [a manchet was basically a roll, a loaf of bread small enough to hold in one hand] and take half a pint of wine and some sugar; the wine must be scalded; then take eight eggs and beat them with rose water, put to them sliced dates, marrow and nutmeg, mix all together and fill the guts to boil.

Gradually, milk and cream started to be added to puddings. One example is A.W.'s Elizabethan "White Pudding of the Hogges Liver":

You must perboile the liver, and beat it in a mortar and then strain it with cream, and put thereto for six yolks of eggs, and grate and put it thereto with a halfe penny loaf of light bread, and put it thereto with small raisins, dates, cloves, mace, sugar, saffron, and the suet of beoffe [beef].

Edward Kidder's calf's-foot pudding from the 1720s stayed in sausage form but included cream:

Take 2 calves feet shred them very fine mix them with a grated penny white loaf being scalded with a pint of cream put to it ½ a pound of shred beef suet and 8 eggs and handful of plump currants, season it with sweet spice & sugar a little sack and orange flower water the marrow of two bones put it in a veal cual [a thin lacy layer of fat around animal organs that is still used to wrap patés and other force-meat dishes] being washed over with the batter of eggs then wett a cloth & put it there in tye it up close when the pott boyles put it in boyle it about 2 hours then turn it in a dish stick on it sliced almonds and citron then pour on it sack verjuice sugar & drawn butter.

Soon pudding could be anything, as long as it included milk or cream, eggs, and sugar. There were barley, millet, oat, and rice puddings; every imaginable type of fruit puddings; chestnut, vermicelli, potato, and spinach puddings. One of the oldest and most popular puddings was hasty pudding. Hasty pudding was made with milk, sugar or molasses, and spices, and it is the same dessert known in New England as Indian pudding, except that the English pudding used flour and the New England pudding cornmeal. There was nothing Indian about the New England dish—the settlers were simply copying a traditional English dish—but cornmeal was an ingredient associated with Indians. Here is one of Amelia Simmons's three recipes for Indian pudding:

3 pints scalded milk to one pint [corn] meal salted; cool, add 2 eggs, 4 ounces butter, sugar or molasses and spice sufficient: it will require two and a half hours baking.

Another very old pudding is plumb pudding. The odd spelling is because the name did not refer to plums, but to assorted dried fruit, especially raisins and currants. This pudding is documented as far back as the fourteenth century, but it may be considerably older. It is very similar to Robert May's wine and guts pudding. Here is Amelia Simmons's recipe for a boiled plumb pudding:

Three pints flour, a little salt, six eggs, one pound plumbs, half pound sugar, pound beef suet, one pint milk; mix the whole together; put it into strong cloth, floured, boil three hours; serve with sweet sauce.

Puddings remained a common feature of large meals into the twentieth century. The Irish writer James Joyce's short story "The Dead" is centered on an extravagant midday goose dinner topped off with a "huge pudding," and as portions were distributed on plates and passed to the many guests down the long table, raspberry or orange jelly or jam was spooned on top. In 1914, when the story was first published in his collection *The Dubliners*, puddings were still in fashion for feasting.

9

EVERYONE'S FAVORITE MILK

HISTORIANS GENERALLY AGREE that *Homo erectus* used fire 600,000 years ago, though some say the practice started only 12,000 years ago and some believe that it began more than a million years ago. But the use of ice, the harnessing of cold, had a much slower development.

There are many theories about why humans were so much quicker to pursue hot than cold. One is that fire is easier to make than ice. But that is not true in a subfreezing climate. Another is that humans were less drawn to cold because heat is associated with life and cold with death.

Natural coldness, such as ice from mountain glaciers and frozen lakes, could easily have been harvested and stored early in human history, at least for those wealthy enough to bear the expense. But for many centuries, humans lived around ice and snow without harvesting it. There was broad debate about the nature of cold, and little understanding of where it came from. One theory held in Europe was that all coldness came from an uncharted island somewhere north of England named Thule. Aristotle believed that the origin of all coldness was water. This theory was not disproved until the seventeenth century, when the chemist Robert Boyle observed that materials that did not contain water could be chilled. He also demonstrated that the coldest part of a body of water was its surface, not its center. That led to the theory that air was the source of coldness, but Boyle also disproved that.

THE EARLIEST EVIDENCE of humans collecting and using ice comes from the town of Mari, which stood on a bank of the Euphrates four thousand years ago in what was then Mesopotamia and is Syria today. Tablets discovered there in 1933 describe an icehouse "which never before had any king built." It also described the ice in the structure as being used up so quickly that though it was used only by aristocrats for chilling wine, servants struggled to keep it stocked.

According to the Old Testament book of Proverbs, "As the cold of snow in the time of harvest, so is a faithful messenger to them that send him: for he refreshes the soul of his masters." Since this was Solomon speaking, it is concluded—by a small stretch—that he knew of the practice of chilling a drink with ice or snow on a hot day. The eleventh-century Chinese also had icehouses. The Egyptians shipped ice from Lebanon.

Early icehouses were not houses but sawdust-lined pits dug in the ground. Certainly the Greeks, Romans, and Arabs knew how to chill their wine with clumps of snow. Pliny is said to have invented the first ice bucket, and in the second century the Romans enjoyed a frozen milk dish they called *mecla*. It is unfortunate that we don't know the exact nature of *mecla*, because this may have been the first frozen dairy dish. But it is also possible that the Chinese made a frozen milk and rice dish a century or more earlier.

A century before this, too, there had been a number of frozen or ice-chilled drinks that were not made with milk. Nero Claudius Caesar sent slaves to the mountains to gather snow that was mixed with fruit, fruit juices, or honey—the first "Italian ices." The Turks also had iced fruit beverages, which they called *sorbet*. Persians called it *sharbate* and Arabs, *sharûb*. These were generally nondairy, but milk did get into sherbet. Antonio Latini, a seventeenth-century chef for

Naples royalty, gave a recipe for *sorbetta di latte* in his 1692 cookbook. Equal amounts of milk and water were mixed with sugar, candied citron, and pumpkin, and frozen. This might have been the first Italian ice cream, though the high water content would make it seem more like a sherbet.

Two favorite myths of food history are about ice cream. One involves Marco Polo and the other Catherine de' Medici, two names that are always red flags for food historians because there are so many erroneous food stories associated with them. The first story has it that Marco Polo brought an ice cream recipe from China to Italy from which the Italians learned ice cream making. The second is that Catherine de' Medici brought Italian ice cream makers to France and taught the French how to make it. Neither of these claims seems likely.

There was ice cream or an ice-cream-like frozen dessert in the Tang Dynasty (618 to 907), and ice cream probably originated in the eighth century, a golden age of Chinese culture. If ice cream truly was invented then, it gets added to an impressive list of Chinese firsts, such as paper, gunpowder, the compass, and printing. But in some accounts, the Chinese learned about the frozen dessert from the Mongolians. Marco Polo may have seen or tasted ice cream while in China in the thirteenth century, but there is no record of an interest in ice cream making in Italy in the fourteenth or even fifteenth century.

In 1533 Catherine de' Medici, a daughter of Italy's most powerful family, was sent to France to marry the Duke of Orleans, destined to become King Henry II of France. Both were fourteen years old at the time. According to the legend, the child bride, wanting to eat well, brought with her a huge entourage of Italian cooks and food specialists. She is said to have introduced not only ice cream to the French, but also the fork, artichokes, and many other Italian items. However, it is significant that she was neither credited nor blamed for Italianizing French food until long after her death, mostly in the nineteenth century.

By Catherine's time, the French and Italians had been exchanging ideas for centuries, and if the Italians had been eating ice cream, the

French would have known about it. Also, in sixteenth-century Florence, where Catherine came from, there were no ice cream makers. But the biggest problem of all, as pointed out by the meticulous British food historian Elizabeth David, was that documents show that the young bride did not bring Italians with her. Her entire entourage consisted of Frenchmen sent to Florence to fetch her.

IN THE SIXTEENTH century, both the French and Italians loved ice and snow, chilled their drinks, and decorated tables with ice sculptures. The Florentines built icehouses in their palaces; the Pitti Palace had snow-cooled wine cellars. In the sixteenth century, a few decades after Catherine moved to France, the Florentine grand duke Francesco was said to drink iced milk with boiled grape must poured over it. Elizabeth David found documents asserting that Francesco also combined frozen milk with egg yolks—which would have resulted in something close to ice cream—but she was unable to confirm that this was true.

All of the leading early ice cream establishments in Europe were started by Italians. And originally, ice cream was only for aristocrats. In the early seventeenth century, Charles I of England, who was said to love his ice cream, would supposedly not allow its recipe passed on to even his noblemen. Of course, this story is probably no more true than the stories involving Marco and Catherine. Like the Medici stories, this one first surfaced in the nineteenth century, long after King Charles I's death. He would also have had to keep his secret extremely well, because there is no record of ice cream in England until Charles II came back from exile in France in 1660.

That same year, Charles II built England's first icehouse, just beyond the eastern wall of what was then St. James's Park. And the first use of the term "ice cream" that can be documented was a reference to its being served to Charles II in 1671.

Ice cream was introduced to Paris by Italians. In 1686 Francesco Procopio dei Coltelli started a restaurant on the rue de l'Ancienne

Comédie called Café Procope. It is Paris's oldest restaurant today. Procopio was one of 250 licensed *limonadiers*, or iced lemonade vendors, in the city. His establishment became the first to sell ice cream to the general public.

Ice cream had probably been introduced to French royalty much earlier, however. Some credit a liqueur distiller named L. Audiger with preparing ices for Louis XIV and his court in the 1660s. And François Massialot, a generation younger than L. Audiger and the chef for Louis XIV's brother Philip II, published cookbooks in 1691 and 1702 (*Le Cuisinier Roïal et Bourgeois* and *Nouvelle Instructions for les confitures, les liquours et les fruits*) in which he included an ice cream recipe—though he called it a cheese, *fromage à l'angloise*. Perhaps he saw a resemblance between his recipe and cheese because he had not learned to churn the mixture while it was freezing and so it came out very dense. Here is his *fromage*:

> Take a chopine of sweet cream and the same of milk, half a pound of powdered sugar, stir in three egg yolks and boil until it becomes a thin gruel; take it from the fire and pour it in your ice mold, and put it in the ice for three hours; and when it is firm, withdraw the mold and warm it a little, in order to more easily turn out your cheese, or else dip your mold for a moment in hot water, then serve it in a compô-tier [a shallow serving bowl].

Though Massialot was certainly not an inventor nor even the first to introduce ice cream to Paris, he did have a famous invention: He was the first to create one of France's most famous dairy desserts, crème brûlée, a custard covered with burnt sugar. He accomplished this by applying a red-hot tool from the fire to the sugar-covered top of the custard. It became so popular that Paris ice cream makers started making crème brûlée ice cream—vanilla with caramel.

Others also called ice cream *fromage*. The celebrated eighteenth-century French chef Menon, who invented the term *nouvelle cuisine*, had a series of ice creams that he called cheeses, *fromages glacés à la crème*. These were poured into molds to resemble various cheeses, including a wedge of Parmesan.

The French of the time had many ice cream specialties as well. Among them was *biscuit de glace*, in which the ice cream was mixed with dried cake crumbs before freezing and then placed on a decorative metal plate or paper napkin. Menon loved *biscuit de glace* because he found the presentation charming.

The most famous ice cream shop in Italy, and now the oldest continuously operated café in the world, was Florian in Piazza San Marco in Venice. It was opened in 1720.

Another Italian, François Xavier Tortoni, opened a café selling ice cream in Paris in 1798. The specialty there, named after him, as was his café, was *tortoni*, made of macaroons, rum, and cream—an elegant variation on Menon's *biscuit de glace*.

In 1750, a Paris café owner known simply as Dubuisson claimed that he was the first to serve sorbets, ices, and ice creams all year round. While he may not have truly been the first, he was certainly one of the first. In the eighteenth and nineteenth centuries, the frozen desserts that had once been exclusively summer fare were gradually appearing in other seasons. Doctors saw these foods as extremely healthful and prescribed them to patients. Given the copious amounts of cream in ice cream and the copious amounts of sugar in the intensely sweet sorbets, most modern doctors would probably not agree. And a century earlier, some doctors would have disagreed as well; it had been believed that iced food and drink caused paralysis. (Conversely, there had also been a time when it was thought iced food and drink could cure paralysis.) Hippocrates warned that eating snow and ice could cause chest colds.

Dubuisson stated that the reason he decided to make his frozen desserts available all year round was because of doctors' prescriptions. Italian doctors in particular insisted on the health benefits of these desserts. Dubuisson quotes a Dr. Mazarini, who claimed that contagious diseases spread more quickly in Italy during years with warm winters and poor snowfalls; cold helped prevent disease, he said.

Another Italian doctor, Filippo Baldini, published a book in 1775 devoted exclusively to *sorbetti*, with each chapter devoted to a single flavor. He gave no recipes, but cinnamon stood out as particularly healthful. He republished his book in 1784 with a new chapter on pineapple, not because of any new health discovery, but simply because pineapple had become a very fashionable flavor.

IT IS NOT known who introduced ice cream to America, but George Washington was known to be a great ice cream enthusiast. Ten lidded ice cream pot freezers were found among his possessions at Mount Vernon after his death. Washington's distant relative, Mary Randolph, published a cookbook in 1824, *The Virginia Housewife*, in which she included six ice cream recipes—vanilla, raspberry, coconut, peach, citron, and almond—along with, true to the period, many more cream and pudding recipes. It is clear that ice cream was made in her slave-owning household, and she offered a great deal of advice on the dessert. "When ice creams are not put into shapes," she wrote, "they should always be served in glasses with handles." She also gave advice on freezing ice cream:

> It is the practice with some indolent cooks, to set the freezer containing the cream, in a tub with ice and salt and put it in the ice house; it will certainly freeze there, but not until the watery particles [have] subsided, and by the separation, destroyed the cream.

Mary Randolph understood that to make ice cream, the custard had to be churned during the freezing process:

> The freezer must be kept constantly in motion during the process and ought to be made of pewter, which is less liable than tin to be worn in holes and spoil the cream by admitting the salt water.

That statement was true enough, but Randolph and her contemporaries did not know of the danger of lead, which is found in pewter.

Here is Randolph's recipe for peach ice cream. The first instruction is the critical secret to this dish:

> Get fine soft peaches perfectly ripe, peel them, take out the stones, and put them in a china bowl; sprinkle some sugar on, and chop them very small with a silver spoon—if the peaches be sufficiently ripe, they will become a smooth pulp; add as much cream or rich milk as you have peaches; put more sugar and freeze it.

Mary Randolph's cousin, Thomas Jefferson (these Virginia Randolphs did not look far for mates; she was also related to both George and Martha Washington and to John Marshall, and while a Randolph herself, was married to another Randolph), was said to have brought back ice cream recipes from France. This is not surprising since he was the original founding American Francophile. He was so partial to French food that Patrick Henry once criticized him for having "abjured his native victuals."

Vanilla was Jefferson's favorite ice cream flavor. He had discovered vanilla beans while in France and brought back two hundred of them, many of which he consumed in ice cream. Among the subjects of his letters to friends in France are pleas to send him more vanilla beans. It is sometimes said that Jefferson introduced vanilla beans to America, but given their popularity in much of Europe, this seems a dubious claim, though he does say in a letter that there were no beans available in America. Jefferson food introductions rank only slightly below those of Marco Polo and Catherine de' Medici in the field of erroneous food mythology.

Jefferson liked to serve ice cream on sponge cake with a lightly baked meringue on top. This dish has endured, with slight variations and different names in America and France, where it was called a Norwegian omelet because Norway is a cold place. In the late nineteenth century, when ruthless, ostentatious millionaires were dining at showy palaces, Charles Ranhofer, formerly of Delmonico's and one of New York's most famous chefs, gave this hopelessly complicated recipe for a simple dish that he called Alaska-Florida because it was both hot and cold. It was one of forty ice cream recipes that he presented in his 1898 book, *The Epicurean*:

> Prepare a very fine vanilla Savoy biscuit paste. Butter some plain molds two and three quarters inches in diameter by one and a half inches in depth; dip them in fecula or flour, and fill two thirds full with the paste. Cook, turn them out, and make an incision all around the bottom. Hollow out the cakes and mask the empty space with apricot marmalade. Have some ice cream molds shaped as shown [cone shaped], fill them halfway with uncooked banana ice cream and halfway with uncooked vanilla ice cream. Freeze, unmold, and lay them in the hollow of the prepared biscuits; keep in a freezing box or cave. Prepare also a meringue with 12 egg whites and one pound of sugar. A few moments before serving place each biscuit with its ice on a small lace paper and cover one after the other with the meringue pushed through a pocket channeled through with a socket [a pastry bag with decorating nozzle] beginning at the bottom and diminishing the thickness until the top is reached. Color this meringue for two minutes in a hot oven, and when a light golden brown, remove and serve at once.

A 1912 book, *Ice Creams, Water Ices, Frozen Puddings, Together with Refreshments for All Social Affairs,* by Sarah Tyson Rorer, a popular Philadelphia food writer, simplified the recipe into a dish that became popular in America. She called it Alaska Bake:

> Make a vanilla ice cream, one or two quarts as the occasion demands. When the ice cream is frozen, pack it in a brick mold. Cover each side of the mold with letter paper and fasten the bottom and lid. Wrap the whole in wax paper and wrap it in salt and ice. Freeze for at least

two hours before serving time. At serving time, make a meringue from the whites of six eggs beaten to a froth; add six tablespoonfuls of sifted powdered sugar and beat until fine and dry. Turn the ice cream from the mold, place it on a serving platter, and stand the platter on a steak board or an ordinary thick plank. Cover the mold with the meringue pressed through a star tub in a pastry bag or spread it all over the ice cream, as you would ice a cake. Decorate the top quickly and dust it thickly with powdered sugar; stand it under the gas burners in a gas broiler or on the grate in a hot wood or coal oven until it is lightly browned, and send it quickly to the table. There is no danger of the ice cream melting if you will protect the underside of the plate. The meringue acts as a nonconductor for the upper part.

Six years later, Fannie Farmer, an administrator and later director of the Boston Cooking School, which tried to simplify cooking for working women, wrote a recipe labeled simply "Baked Alaska."

NEW YORK WAS probably America's first ice cream town. The first ice cream parlor opened there in 1776, and numerous ice cream parlors opened in British-occupied New York during the Revolution. New York not only had ice cream parlors and confectionary shops that sold ice cream, but even had ice cream gardens.

For three years in the 1790s, Jean Anthelme Brillat-Savarin, one of France's first great food writers as well as one of the most enduring, lived in America. He took a great deal of interest in the lives of other French exiles, of which there were many—they had fled the turmoil following the French Revolution. In New York, Brillat-Savarin found a Captain Joseph Collet, who had opened an ice cream parlor and "earned a great deal of money in New York in 1794 and 1795, by making ices and sherbets for the inhabitants of that commercial town." Then he added a sexist comment that was not surprising for a writer who once observed that women were instinctively prone to gourmandism because it was good for their looks: "It was the ladies above all, who could not get enough of a pleasure so new to them as frozen food; nothing was more amusing than to watch the little grimaces they made while savoring it." Sexism aside, it would be amusing to watch the initial response of anyone tasting a frozen dessert for the first time.

In 1797, Brillat-Savarin returned to France, but Captain Collet stayed in New York and opened a boardinghouse, the Commercial Hotel, with a café. In 1835 he sold it to the two Swiss Delmonico brothers, whose famous restaurant had been lost in a downtown fire. In 1824, the Delmonicos—one a wine merchant and the other a pastry maker—established a café in the heart of the business district on Williams Street. It became known as a sophisticated European place, famous not only for ice cream but also for cakes, chocolates, foaming hot chocolate, and Cuban cigars (the wine merchant Delmonico had also been in the Havana tobacco trade). And it was in the café that the New York business lunch was invented.

Around the same time, ice cream was also becoming popular in Philadelphia, most notably at Grays Ferry and Gardens, one of Martha Washington's favorite establishments—she apparently shared her husband's enthusiasm for ice cream. It is sometimes claimed that ice cream was first served in Philadelphia on July 15, 1782, at the French mission, where George Washington was an honored guest. The first known house to sell ice cream to the general public in Philadelphia was founded in 1800 by a Frenchman, Peter Bossu.

In 1818, Eleanor Parkinson opened a confectionary next to her husband's tavern. The confectionary and its ice cream became so successful that the couple eventually closed the tavern. The Parkinsons' café is sometimes credited for giving Philadelphia a reputation for ice cream. Eleanor promoted this idea in her cookbook, in which she stated, "Philadelphia has for a long time enjoyed a pre-eminent reputation in the manufacture of these delicious compounds." "Long time" would be a relative term.

Augustus Jackson, a black cook who had worked in the White House in the 1820s, where he developed improved techniques for making ice cream, opened his own ice cream parlor in Philadelphia, his native town, in 1832. Philadelphia later became famous for its many black-owned ice cream parlors. Jackson's ice cream was said to be the best, and he was particularly celebrated for his variety of flavors. Since ice cream was mainly a summer food, many of the ice cream parlors in Philadelphia, as in New York, were outdoor cafés.

An inglorious moment in American ice cream history involves General "Mad Anthony" Wayne, celebrated for defeating the Shawnee, Lenape, and Miami Indians and driving them off a broad swath of the Midwest to open it for white settlers. After his decisive victory at the Battle of Fallen Timbers, just outside of what is today Toledo, Ohio, he served the troops ice cream, a rare treat that he claimed the army had not "seen since it left the East."

In 1843 the ice cream industry was radically changed by a forty-eight-year-old woman, Nancy Johnson, who invented a hand-cranked ice cream maker. The machine was a wooden bucket filled with ice and an inner metal cylinder, which held the unfrozen ice cream. The handle, which ran through a hole in the bolted-down lid, turned the mixture as it froze. The cylinder even had two compartments, so that it was possible to freeze two flavors at once. The tool became the standard ice cream maker for decades and moved ice cream from the exclusive treat of manor houses with their own icehouses to a popular item. But Johnson, better at engineering than business, never made much money from her invention.

A similar machine, the ice pail maker, was patented in London in 1853 by William Fuller. His patent was later than Johnson's, but he was listed in a directory as an "ice pail maker" in 1842, the year before Johnson's invention. Fuller also had many ice cream recipes, most of them extremely rich in egg yolks, though he did have one recipe that called for only beaten egg whites. In 1843, the same year as Johnson's invention, another Londoner, Thomas Masters, claimed to be the inventor of the first ice cream maker. An advantage of his machine was that it was propped up on a stand so that the user did not need to lean over. Three more ice cream freezers were invented in 1848, one a hand-crank machine designed for professionals to make large batches.

The principle in all these machines was always the same. Ice cream that was frozen without being churned would be dense, hard, and inedible. It was the motion that aerated the cream as it froze, giving it the light texture that makes it ice cream.

In 1844, Masters, the inventor of the London ice pail, published a book titled *The Ice Book: A History of Everything Connected with Ice, with Recipes*. This is his recipe for Howqua's tea ice cream:

> One pint of cream, half a pound of sugar, one ounce of tea, or a suffi-cient quantity to make one cup. Mix with the cream; freeze. One quart.

And ginger ice cream:

> Bruise six ounces of the best preserved ginger in a mortar; add the juice of one lemon, half a pound of sugar, one pint of cream. Mix well, strain through a hair sieve; freeze. One quart.

ICE CREAM IN America became a fast-growing business, especially because it was often more profitable than selling milk. *Godey's Lady's Book*, the most popular women's magazine during the Civil War in both the North and South, regularly ran ice cream recipes. As early as 1850 they wrote that "a party without [ice cream] would be like a breakfast without bread or a dinner without a roast." *Godey's* and most other nineteenth-century recipes usually ended by telling the reader simply to freeze the ice cream. But the freezing was the great trick. This is *Godey's* 1860 recipe for frozen custard:

> Take one quart of milk, five eggs, and a half pound of sugar; beat the eggs and sugar together; boil the milk and pour it on to the eggs

and sugar, beating it at the same time; put it on the fire again, and keep stirring to prevent its burning: soon as it thickens, take it off and strain it through a half sieve, when cool, add the flavor and it is ready for freezing.

And this is their 1862 "pine apple" ice cream recipe:

Pare a ripe, juicy pine apple, chop it up fine and pound to extract the juice. Cover it with sugar and let it lie for a while in a china bowl. When the sugar has entirely melted strain the juice into a quart of good cream, and add a little less than a pound of loaf sugar. Beat up the cream and freeze it in the same manner as common ice cream.

In 1871, in her bestselling cookbook *Common Sense in the Household*, Marion Harland, a Southerner whose real name was Mary Virginia Hawes Terhune, wrote about how to make "Self-Freezing Ice-cream." The recipe makes clear that the really precious ingredient was the ice. It also makes clear that self-freezing, without a crank or similar device, was hard to do:

1 quart rich milk
8 eggs—whites and yolks beaten separately and very light
4 cups sugar
5 pints rich sweet cream
5 teaspoonfuls vanilla or other seasoning, or 1 vanilla bean, broken in two, boiled in the custard, and left in until it is cold.

Heat the milk almost to boiling, beat the yolks light, add the sugar and stir up well. Pour the hot milk to this little by little, beating all the while; put in the frothed whites, and return to the fire—boiling in a pail or sauce pan set within one of hot water. Stir the mixture steadily for about fifteen minutes, or until it is thick boiled custard. Pour into a bowl and set aside to cool. When quite cold beat in the cream. And the flavoring, unless you have used the bean.

Have ready a quantity of ice, cracked in pieces not larger than a pigeon egg—the smaller the better. You can manage this easily by laying a great lump of ice between two folds of coarse sacking or an old carpet, tucking it in snugly and battering it, through the cloth with a sledge hammer or mallet until fine enough. There is no waste of ice nor need you take it in your hands at all—only gather up the corners of the carpet or cloth, and slide as much as you want into the outer vessel. Use an ordinary, old fashion, upright freezer [a metal cylinder] set in a deep pail; pack around it closely; first a layer of pounded ice, then one of rock salt—*common salt will not do.* In this order fill the pail but before covering the freezer lid, remove it carefully so that none of the salt may get in, and with a long wooden ladle or flat stick (I had one made for this purpose) beat the custard as you would batter for five minutes without stay or stint. Replace the lid, pack the ice and salt upon it, putting it down hard on top; cover all with several folds of blanket or carpet, leave it for one hour. Then remove the cover of the freezer when you have wiped it carefully outside. You will find within a thick coating of frozen custard on the bottom and sides. Dislodge this with your ladle, which should be thin at the other end, or with a long carving knife working every particle of it clear. Beat again hard and long until the custard is a smooth, hard congealed paste. The smoothness of the ice cream depends on your action at this juncture. Put on the cover, pack in more ice and salt, and turn off the brine [melted ice and salt]. Spread the double carpet over all once more having buried the freezer out of sight in ice, and leave it for three or four hours. Then if the water has accumulated in such quantity as to buoy up the freezer, pour it off, fill up with ice and salt, but do not open the freezer. In two hours more you may take it from the ice, open it, wrap a towel wrung out in boiling water about the lower part and turn out a solid column of cream, firm, close-grained, and smooth as velvet to the tongue.

This technique of combining salt and ice for freezing is an old one. In 1589, Giambattista della Porta wrote of using saltpeter and ice to freeze wine. A Neapolitan, he was known as the Professor of Secrets, and he gave demonstrations that resembled magic acts. The inquisition in Spanish-ruled Naples questioned him. One of his acts was to chill wine by moving it into what he called a magic elixir. But he revealed the trick along with others in his 1589 book. Salt has a lower freezing point than water and so lowers the freezing point of the ice. This appears to melt the ice, but while the ice turns to liquid, the cold temperature is maintained, creating, in effect, liquid ice, which is much easier to move the ice cream freezer in. Later freezers would use this technique to create a coolant that could move through tubes.

The first book containing only ice cream recipes, a sign that ice cream had arrived, came out in France in 1768. It was written by Monsieur Edy, who did not give his first name, and provided about two hundred pages of ice cream recipes.

THE SODA FOUNTAIN began in Philadelphia at the corner of Sixth and Chestnut streets. Its owner was Elias Durand, a Frenchman who had been an apothecary in Napoleon's army. He had originally intended his place, opened in 1825, to be a pharmacy, but customers were attracted by the sparkling water he sold, and the store became a place to gather and have a glass of soda. People started adding flavors to soda, and a trend for soda and fresh cream began. Then in 1874, Robert M. Green added ice cream to the sparkling water he sold at his store. His counter income went from $6 daily to $600; the ice cream soda had been born.

Philadelphia, a city that liked to brand its food—Philadelphia cheese steak, Philadelphia cream cheese—became even more famous for its ice cream. Sarah Rorer in her 1912 book used the phrase "Philadelphia ice cream." Rorer was one of the founders of the *Ladies' Home Journal* and an influential writer on food and nutrition, as well as the founder of the Philadelphia Cooking School. Her Philadelphia ice cream used no eggs or other thickener, and she insisted on using the freshest ingredients. Or so she said. She sometimes used canned fruit and recommended using canned condensed milk if no high-quality cream was available. To prevent the eggless beaten cream from becoming whipped cream when fresh cream was used, she advised scalding half the cream and, once it was cold again, adding the other uncooked half. She emphasized that eggless Philadelphia ice cream had to be frozen very slowly, and offered many different and often interesting flavors. This one is for apple ice cream:

4 large tart apples
2 quarts cream
½ pound of sugar
1 tablespoonful of lemon juice

Put half the cream and all the sugar over the fire and stir until the sugar is dissolved. When the mixture is perfectly cold, freeze it and add the lemon juices and the apples, pared and grated. Finish the freezing and repack to ripen.

The apples must be pared at the last minute and grated into the cream. If they are grated on a dish and allowed to remain in the air they will turn very dark and spoil the color of the cream.

A MAN NAMED Ed Berners ran an ice cream parlor in Two Rivers, Michigan, and according to this often-repeated and hard-to-believe story, in 1881 a customer named George Hallauer came in and asked for some chocolate sauce on his ice cream. Supposedly no one had ever thought of adding chocolate sauce to ice cream before. In any event, a lot of people around Two Rivers followed suit, and soon there was ice cream with cider, ice cream with chocolate and peanuts (called a "chocolate peany"), and ice cream with many other different toppings. This story gained popularity when H. L. Mencken, one of the most respected figures in the history of American journalism, retold it in his most famous book, *The American Language*. But he told it only as a story he had heard—he never investigated it. Did Ed

Berners, who, according to birth records, would have been eighteen years old in 1881, really own an ice cream parlor?

In the neighboring town of Manitowoc, George Giffy joined in the trend of adding toppings, but, according to lore, did so only on Sundays. When a little girl was refused the treat one day because it wasn't Sunday, she supposedly argued that "this must be Sunday," and got her ice cream, which has been called a "sundae" ever since. No one has yet come up with a story to explain why it was spelled with an *e* instead of a *y*, so we still have that tale to look forward to.

Then again, some claim to already know the story. According to them, religious people were so offended that something as frivolous as ice cream would be called a Sunday that merchants changed the spelling so as not to lose customers. Local historians in Evanston, Illinois, even claim that this alteration in spelling—but not the birthplace of the sundae itself—took place at their local ice cream shop, Garwood's Drugstore.

Ithaca, New York, annoys these Wisconsin towns with its own claim to being the true birthplace of the ice cream sundae, and local historians there have gone to great lengths to document this. As their story goes, on the Sunday afternoon of April 3, 1892, the Reverend John M. Scott went to Platt & Colt Pharmacy, as was his custom after services. He met up with the owner, Chester C. Platt, who was also the church treasurer; this was their traditional Sunday meeting place. Platt ordered two bowls of ice cream from the fountain server, DeForest Christiance, who decided to add cherry syrup and candied cherries to the frozen treat. Scott then dubbed the dish Cherry Sunday.

All kinds of sundaes started to be created, particularly in the Midwest. At the 1934 World's Fair, a sundae was served with hot maple syrup and strawberries. This was said to be the inspiration for Chef Maciel's "Hot Strawberry Sundae" at the Westport room in the Kansas City's Union Station. Here is the recipe:

I pint strawberries, cut in half
4 tbsp Jamaica (dark) rum

¾ cup strained honey
4 tbsp lemon juice
rind of I orange cut in strips
I qt vanilla ice cream

Marinate strawberries in rum one hour. In a small saucepan, slowly bring honey, lemon juice, and orange rind to a boil. Remove the orange rind. Combine strawberry/rum mixture with flavored honey, remove from heat, and serve immediately over vanilla ice cream.

An UNKNOWN ICE cream hawker on the Bowery of New York City is thought to have invented the ice cream sandwich around the turn of the last century. He sold them on the street for a penny each. However, many cultures in many countries have created various kinds of ice cream sandwiches served on everything from local wafers to breads, so it is difficult to say which was first.

No one challenges the 1893 Chicago World's Fair's claim to having invented Alaska Pie, a square of hard frozen ice cream dipped in batter and quickly fried. That may be because no one eats them anymore, but it was the probable forerunner to the Eskimo Pie, invented by Danish immigrant Christian Kent Nelson in Onawa, Iowa, in 1920. He called the chocolate-covered confection an I-Scream Bar. He went into partnership with Russell Stover, the chocolate maker, and Stover renamed it Eskimo Pie. Nelson made a fortune with his Eskimo Pies, claiming to have sold one million a day in 1922. He died a wealthy man in 1992 at the age of ninety-nine.

There are also a number of stories about the ice cream cone. One tells of a man named Italo Marchiony, who immigrated from Italy to New York in 1895 and sold lemon ice and ice cream from a push-cart on Wall Street. At first, he sold his confections in small glasses, but they often broke and had to be constantly washed. So he started baking waffles and folding them while still hot into "edible cups." Soon he had a very successful business, with more than forty pushcarts in his fleet. When he could not make enough edible cups to satisfy the demand, he developed an industrial way to produce them in a factory in Hoboken.

Others believe that the real inventor of ice cream cones was Ernest Hamwi, who sold Persian waffles and introduced ice cream cones at the 1904 St. Louis World's Fair. According to Marchiony's daughter, Jane Marchiony Paretti, her father was selling ice cream at the 1904 Fair when he ran out of edible cups, and Hamwi, the waffle man, pitched in to help him.

This story, widely acknowledged, even by the U.S. Postal Service with a commemorative stamp, has problems. To begin with, seven years before Marchiony immigrated and sixteen years before the St. Louis World's Fair, Agnes Bertha Marshall, popularly known in England as the Queen of Ices for her ice cream and frozen dessert recipes, published *Mrs. A. B. Marshall's Cookery Book*, in which she gave a recipe for "cornets with cream." Very festive-looking cornets, or cones, they were, too. Here is the recipe:

CORNETS WITH CREAM

Mix together into a paste, four ounces of finely chopped almonds, two ounces of fine flour, two ounces of castor sugar, one large raw egg, a pinch of salt, and a tablespoon of orange flour water. Put one or two baking tins into the oven, and when they are quite hot, rub them over with white wax and let the tins get cool; then spread the paste smoothly and thinly over the tins (say one tenth of an inch thick) and bake in the oven for three or four minutes. Take out the tins and quickly stamp out the paste with a plain round cutter about two and a half to three inches in diameter, and immediately wrap these rounds of paste on the outside of cornet tins which have been lightly oiled inside and out, pressing the edges well together so that the paste takes the shape of the cornet; Then remove the paste and slip it inside the tin and put another one of the tins inside the paste so that it is kept in shape between the two tins; Place them in a moderate oven, and let them remain till quite crisp and dry: take them out and remove the tins. These can be kept any length of time in a tin box in a dry place. Ornament the edges with a little royal icing by means of a bag and pipe, and then dip the icing into different coloured sugars; fill them with whipped cream sweetened and flavoured with vanilla, using a forcing bag and pipe for the purpose, and arrange them in a pile on a dish, paper, or napkin. These cornets can also be filled with any cream or water ice, or set custard or fruits, and served for a dinner, luncheon, or supper dish. [If you want to try this recipe—easy to follow, but difficult to carry out—tin cone-shaped molds, "cornet tins," are still available.]

This is the earliest known reference to an edible ice cream cone, so until someone finds an earlier reference, Agnes Bertha Marshall was its true inventor. Her creation took on added importance after 1899, when a germ-conscious London banned the penny lick—ice cream served in a small glass and literally licked clean—because they feared that the glasses were not being properly washed and would spread diseases. In any event, the penny lick was a deceptively shallow glass designed to conceal the true size of the meager one-penny serving.

Marshall became a celebrity in Victorian England, giving lectures and cooking demonstrations to huge crowds. Her specialty was ice cream, and two of her four cookbooks were devoted exclusively to it. She was also a great huckster who put coupons for her books in the drums of the baking powder she sold. She had her own newspaper, which promoted her products, including a wide and shallow ice cream maker; an "ice cave," or insulated box for storing ice cream; and a saccharometer for measuring sugar content.

Her books promoted her products shamelessly. In her seven-point "Hints on making ices," Point 1 is "Too much sugar will prevent the ice from freezing properly" and Point 2 is "Too little sugar will cause the ice to freeze hard and rocky." Both true enough, and a simple way to make it correctly would have been to follow the measurements in

*Agnes Bertha Marshall, "The Queen of
Ices." (ICES: Plain and Fancy, 1885)*

her recipes. But her solution was to use one of the saccharometers that
she sold. Similarly, Point 6 is "Fruit ices will require to be colored
according to the fruit," and she sold a line of "Pure Harmless Vegetable Colors" in bottles with her name embossed on them.

Marshall so popularized ice cream making in England that she
was credited with causing an increase in ice imports from Norway.
However, the true pivotal figure in bringing Norwegian ice to
London was Carlo Gatti, who immigrated to London from the
Swiss Italian town of Ticino in the late 1830s. He started by selling
waffles and coffee from pushcarts but then established permanent
food stands and cafés throughout the city and staffed them with
immigrants that he brought in from Ticino. Gatti introduced London
to the ill-fated penny lick. He also put out many Ticino immigrants
as hokey pokey men.

"Hokey pokey man" was a term used in both London and New
York for Italian immigrant pushcart ice cream vendors. The origin
of the term is uncertain, but is thought to have come from the
mispronunciation of a phrase from an Italian song they sang. The

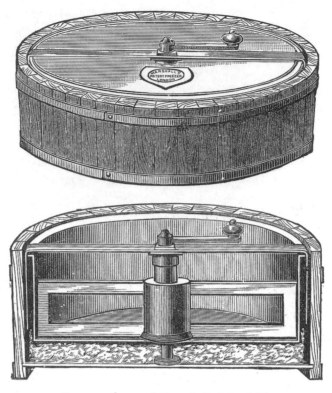

*Ice cream freezer. (Mrs. A. B. Marshall's
Cookery Book,* 1888)

hokey pokey men worked long hours on the streets and on a good
day made one dollar in profit.

The homemade ice creams of the hokey-pokey men were said to
have been made in unsanitary basements and garages, and so the men
were constantly pursued by health inspectors, who claimed they were
spreading disease. Though the accusation was likely true, it was probably also true of many other ice cream vendors. The hokey-pokey
men were likely victims of anti-immigrant sentiment.

In the 1850s Gatti started a major wholesale ice business and he was
the one who started importing thousands of tons of ice from Norway,
which was why ice was readily available for Agnes Marshall's ice
cream trend.

*Agnes Marshall's ice cream molds. (ICES: Plain
and Fancy, 1885)*

In 1904, the year of the World's Fair, Agnes Marshall suffered serious injuries from a horseback riding accident and never recovered, dying the following year at the age of forty-nine. She left behind her four popular books, her cooking school, and a number of machines she had invented, but her family failed to achieve any financial success with any of this, and the Queen of Ices and inventor of the ice cream cone was soon forgotten.

In 1859, THE year before the Civil War began, the total amount of ice cream produced in the United States was estimated to be 4,000 gallons—it was still largely a luxury item, sold to a fortunate few. This estimate is very rough because most ice cream was produced in homes for private consumption. But by 1869, in an expanding and increasingly industrialized nation, 24,000 gallons were produced. And by the last year of the century, 5 million gallons were produced.

The demand for ice cream was far greater than the capacity to produce it. The 5 million gallons sold in 1899 were still made in hand-cranked freezers. However, the problem lay not only in the making of ice cream, but in the storing of it, because without industrial freezers, ice cream makers had to sell whatever they produced within a few hours. Anything left over would go to waste.

Then, at the beginning of the twentieth century, enormous brine circulating freezers became available. These were machines for manufacturers, and allowed for a much greater production of ice cream. In 1904, 12,199,000 gallons of ice cream were produced. Five years later, production had more than doubled, to 29,637,000 gallons. No longer a luxury, now ice cream could be enjoyed by everyone.

Incredibly, however, the wholesale ice cream industry in America began with handmade ice cream before there were freezers in which to keep it. The first wholesale ice cream business was started in the 1850s by Jacob Fussell Jr., a dairyman from Baltimore as well as a Quaker abolitionist and friend of Abraham Lincoln. Like others, Fussell found ice cream to be more profitable than milk—even when selling it at far below the usual prices. His low prices made it extremely difficult for him to keep up with demand, and so he opened an ice cream factory in Baltimore, followed by factories in Washington, D.C., New York, and Boston. With no freezer transportation, his company could sell only in places where they had a plant. Others in the trade constantly accused him of underselling. But he made ample profits with his low-cost ice cream, and others soon followed his example.

Before industrial freezers, ice cream was stored in large iron cans lined with porcelain and placed in cedar tubs packed with ice. But the iron cans could keep ice cream for only a limited time. Delivery wagons were also packed in ice and salt, the salt slowly corroding the wagon parts. Ice cream makers also had to worry about warm winters, which would lead to ice shortages. Ice was harvested in New England and upstate New York and shipped from insulated icehouses there throughout the year.

Fluctuations in the price of salt made pricing difficult, leading some ice cream makers to open their own salt works. Sometimes, too, the ice cream was cheaply made, or too old, and would start to separate, with the top and bottom of a can going bad while the middle was well preserved. Some manufacturers started using egg whites, which helped hold everything together, to prevent this, but sometimes the egg whites made the ice cream too light.

Pellegrino Artusi, a wealthy Florentine silk merchant, whose self-published *L'Arte di Mangiar Bene*, "The Art of Eating Well," has become a classic of Italian cooking, suggested a way of conserving salt when making ice cream. After the ice cream is made, he suggested, boil the brine down to salt crystals and reuse them. Artusi was not extraordinarily frugal, but like many of the wealthy commercial class, he resented taxes, and this was a way around the Italian government's tax on salt.

Artusi's book gave thirteen ice cream recipes. His pistachio called for six egg yolks and his vanilla eight, but he also included many light ice creams made without any eggs. Here is one:

CAFFÈ LATTE GELATO

I quart milk
1½ cups sugar
I pint brewed espresso

Heat the milk and melt the sugar in it, then mix in the coffee and let it cool. Pour the mixture into an ice cream machine, and serve it, in cups or little glasses, when it becomes firm.

A DISTINCTION WAS starting to be made between quality homemade ice cream and an overly whipped industrial product. The *Confectioners Journal* of 1883 called industrialized ice cream "a sham-puffed-up article with no soul or body in it."

The year 1902 marked the beginning of change in the ice cream industry. That summer, the I.X.L. Ice Cream Company of Warren, Pennsylvania, was having trouble getting enough ice. Burr Walker, a son of the company owner, found that a local oil company chilled wax by cooling brine with an ammonia compressor. The principle of cooling is to use energy to rob heat, and compression is a way to vaporize a liquid very quickly. A vapor is expanding liquid. That expansion requires energy, which can be derived by taking heat from objects around it. So the vaporizing ammonia was taking the energy from brine with its low freezing point and chilling it.

Walker made the first brine freezer. Others followed. These were huge cumbersome machines. The Walker freezer had a 12-ton ammonia compressor that made 3 tons of ice daily, which allowed his company to make a thousand gallons of ice cream a day. It circulated 145 gallons of brine at 5 degrees below zero and could freeze ten gallons of ice cream in six to eight minutes.

In 1905 Emery Thompson, a soda fountain manager at a large New York department store, the Siegel-Cooper Company, improved ice cream production by designing a freezer that was vertical instead of horizontal. A new batch of ice cream could be going in the top while a finished batch was coming out the bottom. Because it was vertical, it took up less space—clearly a New York City idea. Working with two machines in a 25-by-60-foot room in the basement of Siegel-Cooper, Thompson produced 400 gallons of ice cream a day.

The new freezers were extremely expensive, and at first only a few companies were willing or able to buy one. But then came the winter of 1915, a catastrophically warm year for natural ice producers, and suddenly the cost of freezers seemed more reasonable—their cost was less than the loss resulting from ice shortages. Soon all wholesale ice cream producers had freezers for ice cream as well as ice-making machines.

The mass-produced ice cream flavors had little in common with the old-time handmade flavors. Vanilla, the most popular flavor, never got near a real bean anymore. The real bean is actually the pod of an orchid that is difficult to cultivate and thus expensive. In the 1870s scientists developed ways of infusing vanilla into alcohol and also started reproducing the flavor from other plants and chemicals. Chocolate, the second most favorite flavor, though still far behind vanilla, was made with cocoa powder—i.e., chocolate with the cocoa butter removed. The number three flavor, strawberry, was made from canned strawberries.

Though largely eaten from April to October, ice cream grew in popularity with a concerted push from the emerging ice cream industry. Ice cream companies sponsored parades to celebrate the opening of the ice cream season in the spring. Sometimes little souvenirs were distributed and sometimes even free ice cream. The opening of the ice cream season corresponded with the opening of the baseball season, and in 1913, the industry let it be widely known that 90 percent of the players on the Detroit Tigers ate ice cream at least once a day and 75 percent ate it at both lunch and dinner.

Freezer trucks were built in the 1920s, and Harry B. Burt Sr., who had a shop in Youngstown, Ohio, started creating ice cream specialties frozen to a stick. He called them "good humor bars," and inspired by the runaway sales of Eskimo Pies, decided to sell his bars from a fleet of freezer trucks manned by a kind of high-technology hokey pokey men.

THE UNITED STATES became *the* ice cream country. By 1919 it was making 100 million gallons annually. Steamer ships started installing freezer compartments so that ice cream could be shipped to India, Japan, and China, despite the often-cited erroneous claim that Asians didn't eat dairy products.

During Prohibition, from 1920 to 1933, with no bars at which to gather, soda fountains became the place for many people to meet. Their popularity tapered off when Prohibition ended.

Freezers were still too large and expensive for retail stores. This did not change until the 1930s, when Clarence Birdseye and General Foods started installing smaller, cheaper freezing units with display windows in stores in their effort to promote frozen food. The final stage of allowing consumers to buy ice cream and keep it at home did not begin until after World War II, when refrigerators started to be built with working freezers.

In Europe, few people had refrigerators in their homes until much later, not even in France. At the beginning of the twentieth century, a French monk, Marcel Audiffren, invented the world's first electric-powered household refrigerator. But he sold the idea to General Electric, and so refrigerators became American.

During World War II, governments saw ice cream as an unnecessary frivolity, a waste of resources. It was banned in Britain and in Italy. The emperor of Japan forced the price below production cost so that no one would make it. Only Americans saw ice cream as a valuable morale booster. The International Association of Dairy Manufacturers and the National Dairy Council, two powerful lobby groups that played a vital role in expanding the popularity of dairy products—too much so, some argued—convinced the U.S. government to include ice cream on their list of essential foods. They did, however, try to restrict the number of flavors and the number of specialty items. The use of excessive sugar, mostly imported, did not fit in with the wartime economy.

The armed forces had ice cream. They made it themselves, and by 1943 they were the world's largest ice cream maker. They built freezer ships to send ice cream to soldiers on the front lines. In 1945 they spent a million dollars on an ice cream ship—an ice cream parlor on a barge.

Certain flavors became American standards, among them peppermint stick ice cream. Peppermint is a hybrid mint bred for its strong taste, and it was extremely popular in both France and the United States. The first peppermint stick was of seventeenth-century German origin, but the red-striped peppermint stick did not make its appearance until the mid-nineteenth century. Soon it was being crushed for peppermint stick ice cream. This 1942 recipe is from the Southern novelist Marjorie Kinnan Rawlings:

I cup boiled custard [beaten egg yolks and sugar gently heated with
   cream until thick] made with half the usual quantity of sugar
IO penny peppermint sticks
I cup rich milk
2 cups Dora's cream [Dora was her cow, known for her unpleasant
temperament and rich heavy cream, which Rawlings claimed was
   what made her ice cream so good]

Crush the peppermint sticks and place with the milk over boiling
water. Stir occasionally until the candy is entirely dissolved [some
peppermint sticks leave a few small crunchy pieces]. Blend with the
custard and chill. Add cream and freeze. This is a lovely pale pink and
has just the right peppermint flavor.

In Europe, especially in France and Italy, but also in the United
States in places such as New England with a strong dairy culture,
small shops continued to make artisanal ice cream. Also, a few
entrepreneurs realized that there was a market for higher quality
ice cream in small containers at high prices. In 1961 Rose Mattus and
her husband, Reuben, developed such a brand and called it Häagen-
Dazs. The success of this brand name proves that Americans like
their food to have foreign names—the way they will use coriander
only when it is called cilantro and the way sherbet has made a come-
back under the name "sorbet." The Mattuses intended "Häagen-
Dazs" to sound Danish, though there is no umlaut in Danish. But
if you wanted a word to look foreign, what could be better than
an umlaut? For almost thirty years, it was the fastest-growing ice
cream company in the world and was sold in twenty-eight countries. But its sale to the Pillsbury Company in the 1980s led to a
series of corporate shifts and mergers, and Häagen-Dazs ended up
a subsidiary of Nestlé.

This has been the fate of most small quality ice cream companies
that tried to grow. In 1978, two New Yorkers, Ben Cohen and Jerry

Greenfield, having completed a correspondence course in ice cream making, opened an ice cream parlor in Burlington, Vermont. Soon they earned a reputation for imaginative flavors with humorous names such as Cherry Garcia, named after Grateful Dead star Jerry Garcia. Cohen, whose ability to taste was limited by an inability to smell, liked ice cream with a lot of texture, and their flavors became known for their chunkiness and textures, such as cookie dough. They were also known for taking environmental stands and, especially, rejecting the use of growth hormones in cattle. They had a huge following, not only for the quality of their ice cream but also for their causes. But in 2000 they sold their company to Unilever.

LIKE WASHINGTON AND Jefferson, the founding father of the Cuban Revolution, Fidel Castro, loved ice cream. His friend the Colombian novelist Gabriel García Márquez recalled in his *A Personal Portrait of Fidel* that the leader once concluded a large lunch by eating eighteen scoops of ice cream. According to CIA documents declassified in 2007, the CIA noticed Castro's ice cream fetish and tried, unsuccessfully, to plant a poison pill in his favorite chocolate milkshake. Apparently the assassin stored the pill in the ice cream freezer and it stuck and fell apart when he tried to take it out.

The Caribbean climate is perfect for eating ice cream but much too hot for making it. Nevertheless, a number of Caribbean islands have a tradition of ice cream making. Jamaican Caroline Sullivan, in her 1893 *The Jamaican Cookery Book*, gave recipes for a number of ices and both banana and coconut ice cream. Here is the banana:

Two bananas
Three eggs
One and a half pints of milk
Sugar

Make a custard of the eggs, milk and sugar to taste. When cold, add the two bananas mashed fine and smooth. Stir, mix thoroughly, and freeze.

But Cuba had a different history than Jamaica. Finally free of Spanish colonialism in 1898, it was taken over by the Americans, who dominated the Cuban economy. Many basics came exclusively from the United States, including gallon containers of the big American commercial ice cream brands. Howard Johnson was popular. Then in 1962 the United States declared a total embargo on imports to Cuba, and the Cubans had to quickly learn how to make many of the things that they had always bought from the United States. Among them were soap, shoes, Coca-Cola, and ice cream.

Fidel Castro took a personal interest in developing Cuban ice cream, and he was determined that Cuba would make better ice cream than the United States. He assigned the task to one of his closest associates, Celia Sánchez, who had successfully brought back the cigar industry after all the leading cigar makers left the island.

Milk delivery in Jamaica, c.1900: This stereograph depicts a woman carrying milk in a can on her head and pouring it into a cup, while a girl waits nearby with her own cup. (Card by Keystone View Company, author's collection)

Sánchez, a fan of ballet, named her proposed ice cream shop Coppelia, after her favorite ballet. The logo was a pair of chubby legs in a tutu and pointe shoes—a warning to ballerinas who ate too much ice cream.

According to legend, Fidel Castro was in possession of excellent recipes for thirty-six flavors of ice cream—some versions of the story said he had more, some less. It is not known where these recipes came from, but given the times, it is usually assumed that they were confiscated. He sent technicians to Canada to learn how to make the flavors and bought top-of-the-line machines from Sweden and the Netherlands. He wanted to build the world's largest ice cream parlor with the world's best ice cream for "the world's best people."

The Cubans did indeed build the world's largest ice cream parlor, as it claimed to serve 4,250 gallons to 35,000 customers a day. Customers had to wait in line for two hours or more, and that line at Coppelia has become part of Havana culture. Originally, Coppelia sold twenty-six flavors, including guava, muscatel, and *crème de vie*, which was an eggnog-like specialty served at Christmastime. Then the country fell on hard times in the 1990s, after the breakup of the Soviet Union. The long lines at the Coppelia have remained, but today's menu board with its original twenty-six slots now has only two or three of those slots filled—usually with vanilla, strawberry, or chocolate. It's a limited choice, but still of high quality. For where would a society be without good ice cream?

ICE CREAM CONES have become not only extremely popular but extremely profitable, far more so than milk, which has a narrow profit margin. Ice cream in general is more profitable than milk, but ice cream cones are one of the more profitable ways to sell ice cream. Alan Reed, a dairy farmer in Idaho Falls near the Wyoming border, has a small shop in which he sells his farm products—fresh milk and cream, his own cheddar cheese, cheese sandwiches, and his own ice cream, sold in cartons or served by the cup or cone. He says his single most profitable item is his ice cream cones. Many retailers would agree.

My own favorite ice cream dish—a childhood favorite that I rarely see anymore unless I make it myself—is *coupe aux marrons*, made with candied chestnuts. This recipe is from Chef Henri Charpentier's 1935 memoir, *Life à la Henri*:

2 pints vanilla ice cream
½ pint sweet cream, whipped
4 tablespoons marrons glacés [candied chestnuts] in syrup. Cut.

Place in bottom of a sherbet glass one tablespoon of marrons glacé, add one scoop of ice cream, surround with whipped cream and decorate with one whole marron.

— PART TWO —

DRINKING DANGEROUSLY

I appeal to you as if you were standing beside a great river in whose current were constantly swept past hundreds of drowning infants.

—Nathan Straus, letter to the National Council of Mayors and Councilmen, September 29, 1897

10

DYING FOR SOME MILK

UNTIL THE END of the seventeenth century, the dangers of drinking milk were often discussed, but the subject lacked a sense of urgency. That changed in Europe and America when the practice known as "artificial feeding," giving babies animal milk in nursing bottles, became common practice.

There is not much literature on artificial feeding, but it had always been a practice, and in a few areas of Europe, such as northern Italy, southern Germany, Iceland, Scandinavia, Switzerland, and Austria— strong dairy cultures—it was very common as far back as the Middle Ages. In some places, too, babies were fed milk with supplements, what would today be called formulas. Babies in Basel, Switzerland, were fed milk with flour and water, and the children there were thought to be healthy. King Louis XV, the long-reigning eighteenth-century French king, had a physician, N. Brouzet, who endorsed artificial feeding. He argued that the children in Iceland and Russia who were artificially fed were extremely healthy, and stronger and less subject to disease than the children in southern countries. In his 1754 book on child healthcare, he wrote that in Muscovy and Iceland, breastfeeding was virtually unknown: "soon after they are born they are left all day, by their mothers, lying on the ground, near a vessel filled with milk or whey, in which is placed a tube, the upper extremity of which the infant knows how to find, and putting his mouth to it sucks, whenever he is oppressed with hunger or thirst." Brouzet said that these infants "escaped the dangers" of infancy better than in France and that "feeding children with the milk of animals certainly is not dangerous."

It is interesting that in eighteenth-century France—and in England, where this book was also popular in translation—artificial feeding was so exotic that it needed to be explained. Artificial feeding was a normal approach not only in Russia, Scandinavia, northern Germany, and Austria, but also in northern Italy, especially the Tyrol. Interestingly, too, in those areas where artificial feeding was dominant, wet nurses were almost nonexistent—and distrusted.

Artificial feeding was not always a choice. In the seventeenth-century American colonies, women were few in number, and the chances of finding a lactating woman available for wet-nursing were slim so artificial feeding was commonplace.

In Europe and America, one way to artificially feed a baby and not risk spoiled milk was to have the child nurse directly from the animal. In sixteenth-century orphanages, especially in France, it was commonplace for babies to suckle on goats. Goats and donkeys were kept for direct feeding in French hospitals, in both the provinces and Paris, into the twentieth century. In 1816 a German, Conrad Zwierlein, set off a trend across Europe of children suckling on goats with his book *The Goat as the Best and Most Agreeable Wet-nurse.* Whether because of their size, temperament, or availability, or because of a belief in the quality of their milk, goats have been used for suckling infants all over the world, from Arabian Bedouins to South African Hottentots. In Europe, pigs were also sometimes used.

In the eighteenth century, scientists learned how to roughly analyze the content of milks, and found donkey's milk to be closest to human milk. The second closest was goat's milk. Donkey's and goat's milk became in great demand for feeding children. However, cow's milk, though less favored, was most commonly used because it was most easily available.

While there have always been some who saw milk as unhealthful, and we now know that huge numbers of people, especially children, have been made sick from milk, there was also a persistent belief that milk was healthful. The fifteenth-century French king Louis XI tried to improve his health by eating more cheese and drinking fresh milk, the latter practice unusual for a wealthy Frenchman of the period. When Francis I of France fell ill in the early sixteenth century, a physician prescribed donkey's milk. He recovered and thereafter drank donkey's milk whenever he was feeling ill.

Cookbooks frequently had sections directed at the elderly and ailing that contained numerous recipes for milk-based remedies. In the eighteenth and nineteenth centuries, "milk water" was a common remedy. This was milk diluted with water with various ingredients added depending on the ailment. Eliza Smith gave two different milk water recipes. This is one:

MILK WATER FOR CANCEROUS BREAST

Take six quarts of new milk, four handfuls of cranes-bill [geranium], four hundred of wood lice; distill this in a cold still with a gentle fire; then take an ounce of crabs-eye [a tropical plant with poisonous seeds] and a half ounce of white sugar candy, both of fine powder; mix them together, and take a drachm of the powder in a quarter of a pint of the milk water in the morning, at twelve at noon, and at night; continue taking this three or four months. It is an excellent medicine.

ONE INDICATION OF milk's standing, at least among the affluent classes, was what was known in France as *la laiterie d'agrément*, the dairy of delight. This was a special miniature dairy designed for the enjoyment of wealthy women. Here, women could milk a cow or two, churn some butter, make their own cheese, and go for a country walk. Visiting *fermes ornées*, ornamental farms, where imitation farming was practiced, was also a popular pastime of the wealthy.

The *laiteries* were perfect little idealized dairies adorned with art depicting mythology and scenes of idyllic nature. The top architects of the day were commissioned to construct them. King Louis XVI had a *laiterie* built in the woods of Rambouillet for his wife, Marie

Antoinette, in June 1786. It featured porcelain bas-relief sculptures of graceful nymphs milking cows and engaged in other dairy tasks. The sculptures were the only dairy images ever produced by the celebrated French porcelain factory at Sèvres.

In that same year, 1786, an ornamental farm at the Trianon of Versailles was also being constructed for Marie Antoinette's amusement. When the king brought her to Trianon in June to present the farm to her, it was at first nowhere in sight. Then a curtain of branches was suddenly thrown back to reveal the surprise.

The queen seemed to harbor a fantasy about the milking life, for she staged a play at Trianon in which she cast herself as a milkmaid and sang:

> *Voilà, voilà, la petite laiterie.*
> *Qui veut acheter de son lait?*
> (Here is the little dairy, who wants to buy some milk?)

The Prince of Wales, the future King George IV (born 1762) churning butter on a farm near Windsor in 1786. From Social Caricature in the Eighteenth Century *by George Paston (pseudonym for Emily Morse Symonds), London, 1905. (HIP/Art Resource, NY)*

How adorable, the king must have thought.

By this time, there was a tradition of French kings building *laiteries* for the women in their family. Louis XIV had commissioned one for his grandson's child bride, the Duchesse de Bourgogne, in 1698. She was said to have milked the cows and made the butter herself that was proudly served at Louis XIV's table.

Like the French aristocracy's gardens, which had become very popular in the eighteenth century, the *laiteries* were places in which to get away and contemplate nature. And there was also the notion, pushed by leading thinkers of the day, including Jean-Jacques Rousseau, that women had a special relationship to dairies. The woman, after all, was a giver of milk.

But alas, Queen Marie Antoinette never used her little dairy. It had not been completed when the king surprised her with it, and shortly after it was finished three years later, a revolution would end both their way of life and their lives.

It is not a coincidence that in the seventeenth and eighteenth centuries, and even as far back as the sixteenth century, wealthy women who thought dairying a pleasant hobby were also mothers who wanted to feed their babies animal milk. This was reflected in their clothing, though there could be a chicken-or-the-egg debate here. Women's clothing had been loose-fitting, making breastfeeding easy. But then the fashion for upper-class clothing changed. Women were wearing tight bodices, dresses that flattened and restricted breasts, clothing not suitable for breastfeeding and perhaps even damaging for lactation. They were being shaped by stiff leather corsets made with whalebone or even metal, laced up the back so tightly that their ribs were occasionally fractured or cracked. And, at the same time, the wealthy started viewing breastfeeding as a lower-class activity.

As UPPER-CLASS WOMEN stopped breastfeeding—and influencing the middle class to do so as well, as the upper class has always had a strong influence on the middle class—increasingly strident voices started to denounce women who did not breastfeed. And, once again, the loudest voices discussing the management of women's bodies were men's. In fact, in the seventeenth and even eighteenth centuries, if a wet nurse was needed, it was the husband who looked for one and negotiated her terms of employment. Bonaventure Fourcroy, a seventeenth-century French lawyer and friend of Molière, suggested that every French home have women breastfeeding their babies under the supervision of their husbands. A 1794 Prussian law required all mothers to breastfeed their children until their husbands ordered weaning.

Cotton Mather, the late-seventeenth-century minister of Boston's Old North Church, a great proponent of witchcraft trials as well as a recognized medical expert who had studied medicine at Harvard, opined that women who did not breastfeed their babies were "dead while they live." Mather, who persuaded Elihu Yale to found a new college because Harvard found his ideas about witchcraft unacceptable, said that God would judge women negatively if they refused to breastfeed. Harvard did not disagree on this point— Harvard president Benjamin Wadsworth termed the decision not to breastfeed "criminal and blame worthy."

According to these men, a woman who did not breastfeed was turning her back on that with which God in his wisdom had provided her. Inevitably, classism entered into the argument, because most of the women who did not breastfeed were upper-class. There was a great deal of talk about women ignoring their maternal responsibilities to live luxuriously. The implication was that they were too idle and vain to live up to the obligations of motherhood. Overlooked in the debate was the fact that women whose labor was essential on family farms or in family businesses needed an alternative to breastfeeding.

The Protestants in particular were advocates of breastfeeding. In fact, attacking women for not breastfeeding was unusual until the Protestant Reformation. Among the more radical sects, such as the New England Puritans, it was rare for a capable mother not to breastfeed. The more radical ministers regularly delivered sermons on the evils of not breastfeeding.

It seemed that men everywhere, both religious and secular, weighed in on what women should do. Benjamin Franklin, master of the homily that says so little yet is hard to refute, proclaimed, "There is no nurse like a mother." Jean-Jacques Rousseau, after abandoning five children to orphanages, expressed strong opinions on appropriate child-rearing and denounced wet-nursing.

At the heart of the debate was a belief, held even before the perils of animal milk were fully understood, that breastfeeding by mothers was the safest option. Another widely held belief was that wet-nursing led to a higher infant mortality rate. Little evidence of this existed among the upper classes, by whom wet nurses were privately employed, but in orphanages, the death rate among wet-nursed infants was horrifying. After visiting French foundling hospitals, many people became passionate advocates for infants being breastfed by their mothers, and some, such as the British physician Hugh Smith, widely read in the British colonies, went so far as to suggest that bottle feeding was safer and preferable to wet-nursing.

Not so, according to Dr. Alphonse Leroy, sent to the foundling hospital in Aix in 1775 to find out why so many children were dying. He concluded that the cause of the deaths was artificial feeding, not wet-nursing. The existence of bacteria was unknown at the time, so he could not have thought that milk could acquire deadly bacteria if not fresh enough. Nor could he have known that the feeding vessels, if not properly washed, could also transmit deadly bacteria. What he did conclude—the right conclusion for the wrong reason— was that the animal milk killed the babies because milk, whether human or animal, became deadly when exposed to air. His successful solution was to have the children suckle on goats.

Still, the infant mortality rate at hospitals remained tragically high, as babies continued to be artificially fed. It was, after all, extremely difficult to maintain a dairy herd at a hospital. At the end of the eighteenth century, a Dublin hospital had a 99.6 percent infant mortality rate. In other words, it was extraordinarily rare for a baby to survive this hospital. It was finally shut down in 1829.

WET NURSES OFTEN came from the lower middle class, from working families. Sometimes they artificially fed their own children so as to have the milk to wet-nurse someone else's. It paid better than being a domestic.

A wet nurse had to have clean habits—no promiscuity or alcoholism—and a happy disposition. Brunettes were most stable. Blondes had cheerful temperaments, but that made them excitable, and that excitement could change the quality of their milk. A study in Berlin in 1838 compared the composition of milk from brunettes, blondes, and redheads and claimed to show definitively that redheads had the worst milk and brunettes the best.

It is curious that for all this concern about wet nurses passing on their characteristics to others' infants, most wet nurses in slave societies were slaves. In fact, slaves were in great demand as wet nurses, so much so that if a slave had a child and was lactating, her sale value would increase.

A baby could also be fed what was called *pap* or *panada*, which meant food supplements—i.e., formula. In poorer regions of the world, additives stretched milk further. In the impoverished cane-growing regions of the Dominican Republic, sugar and water was, and is, a frequent milk substitute. Excavations of Roman gravesites have revealed not only baby milk bottles, indicating a considerable amount of artificial feeding, but also "pap boats" for feeding babies flour and milk or even flour and water.

By the fifteenth century, *pap* had come to mean flour and bread crumbs cooked in milk or water. *Panada* was broth, possibly with vegetables or butter added, cooked with milk and even sometimes eggs. When made with milk, these formulas had an advantage in that the milk was boiled, which made them safer.

Not a great deal is known about Simon de Vallambert, who in 1565 published the first printed book in France on pediatrics. In it he gave this recipe for pap:

> The flour from which it is made nowadays the greater part of the nurses pass simply through a sieve without other preparation. Others cook it in the oven in a leaded or vitrified earthen pot after the bread is drawn, to finally take away the viscosity which is in the crude flour. The milk mixed with the flour is commonly from the goat or cow, that of the goat is better. When one intends to add more nourishment one adds finally an egg yolk, when one wished to guard against constipation one adds honey.

A simpler recipe came from Jane Sharp, an English midwife who in 1671 published the first English book on midwifery by a woman. Her pap was simply "Barley bread steeped a while in water and then boiled in milk."

By the eighteenth century, doctors were starting to endorse cow's milk diluted with supplements, claiming that this made the milk nutritionally closer to human milk. By the nineteenth century, both artificial feeding and milk supplements were widely accepted. In their 1869 book, Catherine Beecher and Harriet Beecher Stowe advised:

> If the child be brought up "by hand" [as opposed to being nursed] the milk of a new milch cow, mixed with one third water and sweetened a little with white sugar, should be the only food given until the teeth come. This is more suitable than any preparation of flour or arrowroot, the nourishment of which is too highly concentrated.

EVEN AS DRINKING milk was becoming more popular and its health benefits extolled, people were becoming increasingly leery of it. One hot Fourth of July in 1850, Zachary Taylor, the twelfth president of the United States, a tough old soldier nicknamed "Old Rough and Ready," laid the cornerstone for the Washington Monument and then refreshed himself with a cool glass of milk. But summer milk is dangerous, and the president died soon after. Many have attributed his death to cholera, but some thought, and perhaps it is true, that it was the glass of milk that killed him.

If the Taylor story is true, he was not the only president to be struck by the ravages of milk. When Abraham Lincoln was seven years old, his family left Kentucky and moved to the small community of Little Pigeon Creek in southern Indiana. In 1818, when he was nine, his mother, Nancy Lincoln, died of something called "milk sickness." It was an epidemic. Nancy's aunt and uncle, and a cousin named Dennis Hanks, also died. Then the disease disappeared for twelve years. When it came back to the area in 1830, the Lincolns left.

Milk sickness was caused when cows ate a plant called white snakeroot, also known as squaw-weed, richweed, pool wort, pool root, white sanicle, Indian sanicle, deer wort, white top, or steria. Known in botany as *Eupatorium urticaefolium*, it usually struck cows, and the humans who drank their milk, in the late summer. In the nineteenth century it decimated communities in what are now the Midwest and Plains states. But incidents occurred even earlier, among the first white settlers in Maryland, North Carolina, Kentucky, Tennessee, Alabama, Missouri, Illinois, Indiana, and Ohio. The disease produced violent vomiting and a burning sensation, and after three days, victims often died. It was often confused with malaria, though the symptoms are not identical.

As far back as in prerevolutionary North Carolina, milk sickness was recognized as a separate disease and was suspected as being caused by milk. Those who abstained from milk, cheese, and all dairy products in the late summer were not stricken, and even those that were, but then abstained from dairy, had only a mild attack.

Some suspected that the disease was caused by a poisonous dew that formed at night. Others suspected that it was caused by an invisible microorganism—one of the early versions of Louis Pasteur's later "germ theory." That was an astute guess, but it actually had nothing to do with the cause of this disease.

Cows grazed on the poisonous white snakeroot plant during late summer and early fall droughts, when the normal grasses were not available and the herds foraged for alternatives. Cows that grazed in enclosed pastures with few weeds did not become infected.

Exactly which weed caused the disease remained a mystery for some time. All the usual poisons were suspected, including poison ivy, water hemlock (*Cicuta maculata*), Indian hachy, Indian tobacco (*Lobelia inflata*), Indian hemp (*Apocynum cannabinum*), Virginia creeper (*Parthenocissus quinquefolia*), cross vine (*Bignonia capreolata*), Indian currant (*Symphoricarpos orbiculatus*), marsh marigold (*Caltha palustris*), spurge (*Euphorbia esula*), mushrooms, and parasitic fungi and molds that grow on various plants. A few people even had it right—white snakeroot (*Eupatorium urticaefolium*). It was a woodland plant, and as settlers cleared more and more land for pastures, the disease disappeared.

BUT IF MILK was killing people in the countryside, it was a small toll compared to the milk-related deaths occurring in cities such as New York, Chicago, and London.

From the first days of colonization, Manhattan stood out as a dairy center because it was settled by the dairy-crazed Dutch. Unlike the English, the Dutch specifically recruited dairy farmers to settle in their colony. Even after the British takeover in 1664, when New Amsterdam became New York, it remained a place where a great deal of dairy was produced and consumed, and the Dutch remained dairy farmers into the next century. Both butter and buttermilk were popular. Bread with butter was a standard breakfast and was also served at dinner. Milk with morsels of bread could be consumed for breakfast or dinner. Even after the British introduced coffee shortly after taking over, the most popular hot drink was tea with milk. New Yorkers also served cheese for both breakfast and dinner.

Even as New York City became more urban, the tradition of owning a cow or two continued. By the nineteenth century, cows were tied to stakes and often fed garbage. Property owners would rent spaces for staked cows and demand a claim on their manure, which sold well to farmers. This did not do much for the smell of the city, but it already had a sewage problem and was a redolent town to begin with.

The old Dutch farms in Europe had been fastidiously kept, but there was no such concern for hygiene in New York City. The cows lived and were milked surrounded by garbage, and the milk was kept in open pails. A street vendor could carry two pails with the help of a yoke across the shoulders and so roamed the streets, ladling out milk to customers.

In the nineteenth century, the Western territories and states with their huge expanses of land became the main producers of grains and other crops in the United States. Easterners could not compete. New England farmland was already showing signs of exhaustion. But the one product with which they could compete, even with limited space, was dairy, and so they became great dairy producers. The extreme example was New York City, which produced tremendous quantities of milk with little space at all.

As transportation improved, milk could be brought to New York City by steamboats that traveled down the Hudson or by train. But the hours of transport on a hot summer day made this milk risky. Cities were probably the worst places for raw milk, but ironically, that was where the drinking of milk first caught on.

MILK DRINKING INCREASED with the growth of cities. It was in cities that milk became the preferred substitute for breastfeeding and the food of choice for weaned toddlers and children.

Milk was supposed to be good for you. There was even a fashion for the "milk cure," six weeks at a milk home drinking six quarts a day. Oddly, though, some of the true health benefits of milk had not yet been discovered. The role of calcium and phosphorus in bone development would not be fully understood until the early twentieth century. But with the industrial revolution and the growth of cities came a belief that breastfeeding was primitive and that the modern, industrialized, urbanized woman was no longer a nutritious milk provider; animal milk was a better alternative.

Old beliefs in the health benefits of whey lingered. Lydia Maria Child, America's first great woman novelist, turned to writing cookbooks when her novels were blackballed because of her abolitionist stand. She and her husband, both antislavery activists, had little money, and she knew how to cook on a tight budget, publishing *The Frugal Housewife* in 1829 and *The Family Nurse* in 1837. In the later book she gives recipes for nine different wheys made by adding acid to fresh milk—vinegar, orange, cider, wine, lemon, and others. All were offered as cures for various complaints: lemon whey for high fevers, molasses whey for wet nurses with insufficient milk, mustard whey for low fevers and nervous fevers, etc.

Yet as the demand for animal milk in the cities increased, its quality grew worse. A few cows staked here and there could no longer provide enough milk for the many clamoring customers. Large stables holding hundreds of cows were established adjacent to breweries, and milk became a big, profitable business. The leftovers from making beer—the mash, or slop, or swill—was poured down wooden chutes into dairies next door. But beer waste was not good feed for cows, and the milk they produced was low in fat and watery, with a light blue color. Producers added annatto to improve its color and chalk to give it body. They also added water to increase the amount of milk they could sell, and covered up the dilution by adding more chalk. Sometimes a little molasses was added, too, to give the concoction the slightly sweet flavor of fresh milk.

By the 1840s almost half the babies born in Manhattan were dying in infancy, mostly from cholera. There were many theories as to the cause of this high infant mortality rate, but it was Robert Milham Hartley, a temperance crusader, who was the first to put the blame on the milk produced in the brewery dairies.

Hartley was a social reformer with many causes. As a young man, he had left his job as a factory manager in the Mohawk Valley and moved to New York City to agitate for a variety of social issues, which he continued doing all his life. Temperance and the plight of the poor were his leading concerns for a period. But then he took on milk, which he believed to be a perfect food, as least when produced by wholesome methods. He was probably the inventor of the term "swill milk," whose production he wanted to expose and stop. He reported that "about ten thousand cows in the city of New York and neighborhood are most inhumanely condemned to subsist on the residuum or slush of this grain, after it has undergone a chemical change, and reeking hot from distilleries."

Hartley also reported that the crowded brewery stables were filthy and that many of the cows were sick and dying, but being milked nonetheless—sometimes even when they were too weak to stand and had to be held up by straps. He identified five hundred dairies in Manhattan and Brooklyn, most around the edges of the city, producing 5 million gallons of doctored bluish milk a year. Many were located near the Hudson River or between Fifteenth and Sixteenth streets, which was then the northern edge of the city. He reported that an unbearable stench came from the stables, as they had no cleaning facilities and no ventilation. He also noted that numerous European countries, notably England and Germany, had brewery dairies too, and that swill milk was being made in Boston, Cincinnati, and Philadelphia as well.

But Hartley's most important point was the possible connection between the swill milk and the rise in infant mortality. In 1815 children under five had represented 33 percent of deaths in Boston, which was horrific enough, but by 1839, children under five represented 43 percent of Boston deaths. Children under five had represented 25 percent of deaths in Philadelphia and 32 percent in New York in 1815, but rose to over 50 percent in both cities in 1839. The rapid rise appeared to correspond with the rapid growth of milk production in brewery dairies. Was this milk poison?

It is not certain the extent of the impact of Harley's 1842 book *An Essay on Milk*, but it was first to raise the issue of swill milk, and started a debate on the subject that took fifteen years to fully erupt. In 1848 the New York Academy of Medicine studied swill milk and concluded that it had far less nutritional value than milk from farms. This was a significant finding, as a fair number of infant deaths were caused by malnutrition.

There was also another huge problem with the milk: microorganisms. But until the mid-nineteenth century, almost nothing was known of these invisible organisms, and it would take another forty years before their ability to spread disease was fully understood.

By 1855, the 700,000 New Yorkers living in the largest city in the United States were spending $6 million annually on milk. More than two thirds of that was spent on swill milk, and infant mortality was continuing to rise. Between the time Hartley's book was published in 1842 and 1856, the percentage of children under five who were dying in a year had more than tripled. More and more people began to wonder if that was related to swill milk.

In 1857 the Common Council of Brooklyn decided to investigate and issued a report that shocked New Yorkers who drank milk or fed it to their children. The report was also difficult reading for anyone who had any regard for the treatment of animals. The report described how cows were brought to the dairy, tied in one spot, and kept there for the rest of their lives. Steaming brewery slop flowed past them in a trough three times a day while they stood in their filth and waited for the slop to cool enough to eat. On average a cow ate thirty-two gallons of this slop a day, but was given no water because it was thought that the swill contained all the water the cows needed. Since they were given no solid food on which to chew, they often lost their teeth.

The Brooklyn report spurred the very popular magazine *Frank Leslie's Illustrated Newspaper* to take on the issue with a series of devastating articles, accompanied by beautifully drawn illustrations such as one of a suspended cow, nearly dead, being milked—cows fed on swill might not have been healthy, but they did produce a lot of milk. A May 1858 article in *Leslie's* stated, "Swill milk should be branded with the word 'poison' just as narcotics are." In the summer, the temperature in the enclosed stables could be as high as 110 degrees Fahrenheit. The manure piled up around the cows, and on those rare occasions when it was shoveled out, it was thrown into a nearby river. Cows in distillery dairies usually lived for only six months.

In New York in the 1850s, cows were fed swill that contained residue from nearby distilleries. The milk these cows produced became known as swill milk and resulted in a major adulterated food scandal as it killed thousands of infants in a single year. (Harper's Weekly, August 17, 1878)

In addition, the cows frequently came down with tuberculosis, but still they were milked. Tuberculosis in cows can infect humans who ingest their milk, but at the time it was widely believed that bovine tuberculosis could not be transmitted to humans. Thus as late as 1913, few Londoners were upset when a test showed that one in ten milk samples arriving at the London train stations contained tubercle bacilli.

Swill milk was loaded onto carts and sold on the city streets. Often the cart had a sign saying PURE COUNTRY MILK or GRASS FED MILK. This was not milk for the poorest people—they could not afford it and breastfed their babies. It was milk for the working and middle class, and even some of the affluent. But by the 1860s, the word was out. In her 1869 book, Catherine Beecher warned that if a child was reacting badly to milk, a parent should first ascertain if the milk really came from "a new milch cow . . . as it may otherwise be too old. Learn also if the cow lives on proper food. Cows that are fed on still-slops, as is often the case in cities, furnish milk that is often unhealthful."

Because of the *Leslie's* campaign, a number of brewery dairies closed. Others were cleaned up, and in the second half of the nineteenth century, under public pressure, milk purity laws were passed and brewery dairies closed. Late in the century, the lactometer was invented. It could measure the amount of solids and fats in milk, and made state laws on milk purity enforceable. In the state of New York, milk was required by law to be 12 percent solids, of which at least 3 percent had to be milk fat. If not, the producer was fined. The irony is that some of the bestselling milks today—0 percent, 1 percent, and 2 percent fat—were illegal in the nineteenth century. But the perception of fat has shifted. People today tend to view it as unhealthful, something to avoid, whereas it used to be a sign of quality.

Even after the lactometer was invented, milk, including true "pure country milk," still killed people at times, especially children. There was a scientist in France who had a theory. But few believed him.

11

THE FIRST SAFE MILK

A<small>N INTERESTING BREAKTHROUGH</small> regarding milk was reported by Phineas Thornton of Camden, South Carolina, in his 1845 *The Southern Gardener and Receipt Book*:

> A foreign journal states, that some milk was lately exhibited in Liverpool from on board a Swedish vessel, that was several months old, having made two voyages from Sweden to the West Indies and back again, and remained perfectly sweet and fresh.

He went on to describe a newly discovered industrial process—canning—that had made this possible, and added:

> It is evident this discovery will be most available at sea; but when bottles could be easily obtained, many families living in cities and villages who keep a cow, might, by preserving some in this way, furnish themselves with a supply for the time a cow usually goes dry for the winter. In any event the experiment could cost but little.

Canning was one of the first food inventions of the Industrial Revolution. Like most of the early industrial inventions, it was a French idea that the British were first to develop. The French had scientists and engineers, but the British had entrepreneurs. When

Napoleon was sending armies around the world, portable food that would not spoil was a great challenge for the French military, and a 12,000-franc prize was offered for a good solution.

Nicolas Appert, a chef and candy and liqueur maker, spent fourteen years developing his response: Food, if thoroughly sealed in a glass jar and heated, would not spoil. He experimented with vegetables, stews, fruits, jam, and also sterilized milk. But his milk experiment was not entirely successful because the resultant product had an unpleasant taste. He wrote a book on his method and it was translated into English in 1809. As soon as it appeared, a Londoner named Peter Durand patented the exact same idea. And had an additional thought: Why did glass jars have to be used? Perhaps other types of containers would work better. Not long thereafter, a man named Bryan Donkin built the first canned food factory on the Thames.

But milk wasn't being canned. That would come later. Rather, it was being put up in jars like jams and preserves. It is not known if the Swedes were the first to apply canning to milk. But they may have been, since this was the country that consumed the most milk per capita. Those who point this out, next start talking about how tall and healthy they were.

CANNED FOOD ARRIVED in the United States in 1819, but was not popular until the Civil War, when its usage was spurred either by military necessity or by the discovery that the addition of a salt, calcium chloride, to the water raised its temperature and made the process more efficient.

At the same time, a certain amount of interest in putting up food in jars arose, and recipes for this process began to be included in cookbooks. Among them were recipes for preserving milk.

In 1867, Annabella P. Hill, a Georgia judge's widow with no particular scientific background, wrote *Mrs. Hill's Southern Practical Cookery and Receipt Book*, which went through numerous editions and was very influential in the second half of the nineteenth century. Her interest in wholesome milk may have been influenced by the fact that five of her eleven children had died before the age of five—not unusual at the time. She gave this recipe, "To preserve milk for a journey":

> Put the fresh sweet milk into bottles; put them into an oven of cold water; gradually raise it to the boiling point; take them out and cork immediately; return the bottles to the boiling point; let the bottles remain a minute. Take the oven from the fire, and let the bottles cool in it.

ANOTHER PROBLEM WITH nineteenth-century milk, aside from merchants who deliberately diluted it, was all of the dirt, twigs, leaves, and other detritus that fell into the milk pails by accident. The fact that an open pail was considered an acceptable conveyance for fresh milk indicates how little thought was given to hygiene.

According to the legend, in 1883 Dr. Henry G. Thatcher of Potsdam, New York, was standing in line buying milk. In front of him was a little girl with a very dirty well-worn rag doll. While the vendor was scooping up milk from his bucket to fill her pitcher, the little girl accidentally dropped her doll in the bucket. But the kind milk vendor averted the crisis by fishing the doll out, shaking it off, and handing it back to the girl. Then he served Dr. Thatcher his milk.

Whether or not there actually was a doll, this tale, like all good stories, illustrates a truth. The merchant was untroubled by the doll's falling into his milk, nor did he expect it to bother his customers.

The supposed event led Dr. Thatcher to patent a milk bottle with a sealable lid a year later. It wasn't much of an invention, considering that Appert had started putting up milk in sealed jars eighty years earlier. Yet it was the first milk bottle and a huge step toward safer milk.

Not everyone in the dairy industry was happy about the new idea. Now they would have to buy these bottles and replace them when they broke, which would probably be a frequent occurrence. And the health authorities would probably demand that the bottles be thoroughly washed after each use. But consumers were pleased with the idea of getting their milk in sealed bottles rather than ladled from dirty buckets, and by the turn of the century, most milk came in bottles. Farmers delivered milk to dairies, and dairies bottled the milk at a plant. The dairy business was looking less like a family-run operation and more like an industry.

After the acceptance of bottles, the term "artificial feeding," which had been popular earlier, gradually fell out of favor. Now the term to use was "bottle-feeding."

Bottles helped promote the idea of adding other ingredients to milk—of making "formula." In the 1860s, many doctors and household guides advised mixing milk with various combinations of water, cream, and honey. In 1867, a German pharmacist in Switzerland, Henri Nestlé, offered his neighbor a mixture of fresh milk, wheat flour, and sugar for his ailing child. The child recovered, and, being a good entrepreneur, Nestlé bottled his formula and claimed to have saved the child's life. In some versions of this story there was no sickly neighbor. But either way, Nestlé's formula involved adding ingredients to milk; he called his invention a good "Swiss milk and bread." It was the world's first commercially sold bottled infant formula and the beginning of the Nestlé company in Vevey, Switzerland.

Formula was based on the observation that human milk appeared thinner and tasted sweeter than cow's milk. So to make cow's milk more closely resemble human milk, it had to be watered down and sweetened. But then it was observed that this formula lacked the fat content of human milk. A little cream was added. Some, observing that human milk was alkaline and cow's milk more acidic, suggested adding water to correct the acidity level. Really, everyone was just painting in the dark—guessing how to make the artificial equivalent of human milk out of cow's milk.

Then in 1884, A. V. Meigs, a Philadelphia doctor, published his chemical analysis of human milk, which became the standard. His laboratory used what were for 1884 extremely sophisticated techniques, and he concluded that human milk was 87.1 percent water, 4.2 percent fat, 7.4 percent sugar, 1 percent inorganic matter such as salt and ash, and 1 percent casein, which is protein. He then analyzed cow's milk and found that it contained 88 percent water, 4 percent fat, 5 percent sugar, 0.4 percent ash, and 3 percent casein. So the early makers of formula had been wrong to add water, and right to add fat and sugar.

Meigs was concerned about the higher casein level in cow's milk compared to human milk. Casein coagulated firmly and Meigs believed that too much of it was indigestible for babies. He therefore recommended adding lime water to cow's milk to both break down its casein and make it alkaline. To adjust its sugar level, he recommended adding more lactose, the sugar already in milk, and increasing its fat content by adding cream. This became the formula that was used for years. Its inescapable flaw is that not all human milk is the same. Not all cow's milk is the same, either. For example, the milk of a Jersey cow has more fat than the milk of a Holstein. Still, it was a formula that people trusted.

Formulas convinced many women, at least those who could afford it, that there was a suitable substitute for breast milk. A study in the United States at the end of the nineteenth century showed that 90 percent of working-class women still breastfed, but only 17 percent of middle- and upper-class women did. In the twentieth century, with the advent of safe pasteurized milk and improved commercial formulas, breastfeeding declined even further. By 1950, more than half of all American babies were fed formula. But this increase was also thanks to another nineteenth-century invention: canned evaporated milk.

IN 1828, AMERICA's first commercial canner, William Underwood, preserved milk in a bottle with sugar, but it didn't sell. In 1847, a Belgian named Francis Bernard Bekaert improved on the formula by adding carbonate of soda. That same year, Jules Jean Baptiste Martin de Lignac obtained a patent for a process that evaporated milk to one-sixth its volume with a little added sugar. But all these plans failed because the milk fat separated and did not sit well in the liquid. The milk tasted overcooked, even burned. It was an unappealing product.

Today, Gail Borden is remembered as the inventor of condensed or evaporated milk. He used the word "condensed," but "evaporated" is an often-used alternative. The milk was condensed with an evaporator—Borden really didn't invent anything. But he was the first to produce preserved milk that was appealing. Often the inventor that history remembers is not the true inventor, but the one who made the idea commercially successful. Thomas Edison didn't invent the lightbulb either.

Gail Borden Jr., born in 1801, was from an American establishment family, a descendant of Rhode Island founder Roger Williams and two signers of the Declaration of Independence. Although he had less than two years of formal education, he became a surveyor and a newspaper publisher. By the 1840s, the Industrial Revolution was beginning to explode with life-changing new ideas, and he, like many others, decided to become an inventor. At this point living in Galveston, Texas, his first invention was the "locomotive bathhouse," a closed room on wheels for women to wade into the Gulf of Mexico safe from sun and surf and prying eyes. Then he built a kind of wagon, a prairie schooner, or, as he labeled it, "terraqueous machine, that had sails and could also do water crossings."

Next he became interested in industrializing food. His first idea was to construct a large-scale refrigeration facility. Then in November 1846 a group of eighty-seven people led by George Donner and James Reed became trapped in heavy snow in the high Sierra Nevada. They were not rescued until February, by which time only forty-eight were still alive. Many had died from starvation, and many of the survivors had stayed alive by eating the dead. This was a well-known and sensational story in 1847, and Gail Borden kept thinking that everyone in the Donner party could have survived if they had had well-preserved provisions.

The invention that Borden then came up with was dehydrated meat biscuits. Meat was dried in an oven, mixed with flour or vegetable meal, and pressed into thick crackers. He imagined huge sales orders coming in from armies all over the world and from explorers and migrants making long treks, like the Donner party. The biscuits won him a gold medal at a London exhibition, but no one wanted to buy them because it was generally agreed that they tasted awful.

On Borden's return voyage from London, his ship encountered rough waters, and the two cows kept in the hold to provide milk to the infants onboard became too distressed to do so. Several of the infants died. This apparently greatly troubled Borden, because when he got back to the United States, he started working on a way to preserve milk in cans.

Borden's first attempt, boiling milk in an open pan with some molasses added, kept very well, but was a dark color that was said to be ugly. People also disliked its molasses smell.

Advertisement for Borden's condensed milk, c. 1888.

In 1853 Borden went to a Shaker community in New Lebanon, New York, to look at an interesting piece of equipment they had called a vacuum pan. It had been invented by an Englishman, Edward Charles Howard, in 1813 for the refining of sugar. The pan reduced the pressure on a boiling liquid to below the pressure normally exerted by its escaping vapor, and this enabled the liquid to be evaporated at a far lower temperature.

Eighteen years before Borden's visit to New Lebanon, another Englishman, William Newton, believing that evaporated milk would taste better if it were evaporated at a lower temperature, had been the first to use the vacuum pan for milk. But he had never tried to market his idea.

Borden's first attempts did not go well. The milk stuck to the sides of the copper evaporation pan. He decided to try greasing the pan, and this worked much better. The milk tasted good. But the patent office at first rejected his invention, saying that it was not an invention at all—it had all been done before. This was somewhat true, and yet Borden had produced a better condensed milk than had ever been made before. He kept submitting his patent application, and on his fourth try, in 1856, he was finally awarded a patent for condensed milk made in a vacuum pan with the addition of sugar. Borden's "sweetened condensed milk" was put on the market in 1860, just in time to sell to the fast-growing Union Army.

He also went to market just as Frank Leslie's campaign against swill milk was scaring everyone in New York off milk direct from the cow. Borden, on the other hand, offered New Yorkers milk in a can for their babies that was both safe and sweet.

12

A New and Endless Fight

THOSE FOR WHOM it has seemed odd that the French, who have had so little interest in drinking milk, could have such an impact on milk production can take comfort in the fact that Louis Pasteur was not particularly interested in milk. His concern and his research were primarily focused on beer and wine. But his idea, his "germ theory"—so called because it took time before it was accepted as fact—had a huge impact on dairies and on public health and medicine in general.

Easy to state but complicated to demonstrate, Pasteur's theory was that there are tiny organisms, invisible to the naked eye, that cause disease—and other effects, such as fermentation. Some germs are useful and some are harmful. Pasteur's theory explained a number of things about milk that were already known. It explained why people got sick and died from milk, why unhygienic dairies were more likely to cause illness, and why fermented milk products, such as cheese and yogurt, tended not to make people ill, not even in warm weather.

Since for centuries it had been understood that warm weather turned milk, farms were equipped with "spring houses," rooms that were cooled by the constant flow of cold water from wells or springs. Since it was also believed that milk would spoil in a thunderstorm— perhaps because of the lightning—it was often kept in containers made of a nonconductive material such as glass.

Consumers had long been provided with household tips on how to judge milk. Here is Elena Molokhovets's advice:

1926 French 90-centime Louis Pasteur stamp.

Good milk is somewhat heavier than water, because drops of
milk sink in water. If whole, good milk, is dropped on a
fingernail, the round drops will hold their shape, but milk
diluted with water will spread out. Good milk is thick and
pure white, but adulterated milk is thin and falls in a bluish
tint.

Rub some between your fingers to determine whether or
not it is fatty.

She also pointed out that boiling milk, the leading way of puri-
fying it from possible diseases, robbed milk of its nutrients.

The existence of bacteria had been known since the late seventeenth century, when the Dutchman Anton van Leeuwenhoek built
a microscope that magnified subjects 270 times, as opposed to
previous devices that at best magnified 50 times. With this tool he
could see many little creatures, bacteria, squirming in a drop of water.
But until Pasteur, there was no solid understanding of what bacteria
did. Even the word "bacteria" did not exist until the German
naturalist Christian Gottfried Ehrenberg came up with it in 1838.
Scientists were also only beginning to understand that bacteria
existed not only in water, but everywhere.

Pasteur's theory held that some of these bacteria, called "germs,"
caused disease, but he couldn't prove it. At the time, it was widely

believed that diseases were caused by miasmas, vapors that rose from the earth. In 1854, John Snow, an English physician who championed hygiene in medicine as well as the use of anesthesia, claimed that cholera was spread through germs that lived in unclean water. But few believed him.

William Budd, an English country doctor, the son of a doctor, and one of seven out of ten siblings who were doctors, from 1857 to 1860 published a series of articles in the *Lancet* showing that typhoid was not spread by bad air but by contagion, from person to person. Though many rejected his findings, he continued to study typhoid epidemics and found much evidence that supported his conclusions. A teacher in the Bristol Medical School, he also advocated for the use of disinfectants, or germ killers, and by the time he published *Typhoid Fever* in 1874, he had significantly altered the medical approach to epidemics. A 1849 cholera epidemic in Bristol had killed two thousand people. In 1866 another cholera epidemic hit the city, and this time, thanks to the implementation of Budd's ideas, only twenty-nine people died.

Robert Koch, a German Nobel Prize—winning scientist who taught himself to read at the age of five by studying newspapers, brought more new ideas to medicine and milk. In 1860 he studied anatomy at the University of Göttingen under Jacob Henle, a disciple of the new "germ theory." Koch then examined the spread of anthrax, an infection caused by a spore-forming bacterium, while he was serving in the German Army in the Franco-Prussian War. With no equipment, he used wooden splinters to inject the anthrax-causing bacteria into mice (ouch!). He went on to work on the spread of other diseases, and in 1882 he concluded that there were three distinct tuberculosis germs: a rare one spread by birds, a common one spread from one person to another, and a third one, not as rare as the first nor as common as the second, spread through milk.

This 1882 discovery, really proof of a belief that had existed for some time, changed the dairy industry. It led to what really should have become known as the "kochization of milk," not the "pasteurization of milk," as we call it today. But Pasteur was the one who identified the process to eliminate the disease that Koch had later discovered.

Bovine tuberculosis, a disease found in cattle, is transmitted to humans through milk. It attacks the glands, intestines, and bones. Humans who survive the disease often become hunchbacked or deformed in other ways. Children are particularly susceptible and are often kept in braces for years to keep their spines from becoming deformed.

Research by the British army in Malta in the 1880s and 1890s led to the discovery of another germ that could be transmitted to humans through cow, sheep, or goat's milk. It caused what was called Mediterranean fever, whose symptoms were severe joint pain, sweats and chills, and fevers that lasted as long as six months. On occasion, the symptoms were permanent. The bacterium that caused the disease was named Brucellosis, after the British doctor David Bruce, though historians believe that the discovery of the disease should be credited to the Maltese doctor under him, Themistocles Zammit.

A similar discovery, bad news for goat's-milk advocates, was that raw goat's milk can carry a bacterium called *Brucella melitensis*, which causes undulant fever, heavy perspiration, and aching joints, a condition that can last for weeks or months.

The more scientists investigated, the more milk-borne diseases they found. Serious intestinal diseases could be transmitted through milk from unclean udders. Farmworkers with contagious diseases could transmit them through the milk pail. Scarlet fever, diphtheria, and typhoid were all traced to contaminated milk.

Milk then became the first laboratory-tested food. In 1887 the U.S. Public Health Service, a government agency founded in 1870, established a laboratory for that purpose.

In 1892 the United States started testing all dairy herds for bovine tuberculosis. The test had been developed accidentally by Robert Koch while he was trying to develop a vaccine against the disease. His vaccine hadn't worked, but the cows that were injected with it developed an inflammation at the injection point if they had the disease. It became a way of testing for the tuberculosis bacillum, and the results were horrifying. A significant portion of American cows, and by extension American milk, was infected. The infected cows were then removed from the herds, and the number of cases of bovine tuberculosis in humans dramatically declined.

In the 1880s the idea of sterilized milk came to the United States. It was believed that the lives of babies could be saved if they were fed milk that had been boiled and then cooled. Louis Pasteur had developed the sterilization process in France in the 1850s and 1860s. As a passionate believer in science serving industry, he had taken a professorship in Lille in northern France, a region of distilleries, in 1854. He had wanted to learn why liquids soured and spoiled, and so resolved to examine all substances that fermented to see if they all contained living organisms. He started with milk because he thought it would be the simplest to demonstrate. It was more difficult than he had imagined, but he did manage to show that lactic acid fermentation was caused by living organisms.

Still only thirty-five years old, Pasteur then left milk to study other substances and the larger questions of where these organisms came from and how to get rid of them. He discovered that wine that had gone sour contained active living organisms, but if heated to between 140 and 158 degrees Fahrenheit (far cooler than the 212 degrees at which water boils) and kept at that temperature for a few minutes and then rapidly cooled, the wine never soured. This was the original pasteurization process. Pasteur went on from there to work on many other projects, including research in the field of immunization against diseases such as anthrax, cholera, and rabies.

By the time the word "pasteurize" was applied to milk, Pasteur was in the final years of his life; he died in 1895 at the age of seventy-two. He had developed the pasteurization process in 1864, but it took decades before scientists began to apply it to milk. When they did, they found that if they heated milk for twenty minutes to just below its boiling point and then rapidly cooled it, it would neither sour nor carry diseases. The process also killed the good bacteria, however, which is why many cheesemakers refuse to use pasteurized milk. Some consumers then and now complain that pasteurized milk is dead. But it was also argued that boiling destroyed everything, but pasteurization, because it was below the boiling point, allowed nutritive elements to live.

With the new science there were two possible approaches to milk as a public health problem. Government could require the pasteurization of all milk, despite the fact that many people disliked it, or they could set up a system of inspection to guarantee the quality of raw milk, which was called "certified milk." Henry Coit, a Newark, New Jersey, physician, established a network of panels of physicians, known as a medical milk commission. The first bottle of certified milk was produced by the Fairfield Dairy in 1894. In 1907, commissions from around the country had banded together to found the American Association of Medical Milk Commissions. A certificate from a commission earned the right to label the milk as certified milk, which would fetch a far higher price. But as would happen with other special-quality milks in the next century, the consumer was not willing to pay a high enough price to cover the added cost of certification.

Meanwhile, in cities such as New York, Boston, Philadelphia, and Chicago, infant mortality remained high for the next twenty years while the merits of pasteurization and certified milk were debated.

NATHAN STRAUS, BORN in Germany in 1848, emigrated to a small town in Georgia in 1856 with his two brothers and mother to join his father, who had moved there two years earlier. But the Straus family lost most of their money during the Civil War, and in 1865 they moved to New York, where the Straus brothers eventually gained control of two of New York's largest department stores, Macy's and Abraham and Straus.

Nathan Straus was guided by a social conscience. He provided low-cost lunches and health care to his employees. In the harsh winter of 1892–93 he distributed coal to the poor and provided affordable lodging for the homeless. He was concerned about the high infant mortality rate in New York City and became convinced that it was caused by milk and that the solution was pasteurization. In June 1893 he opened the first of what he called "milk depots" on the East Third Street pier in the impoverished immigrant area of the Lower East Side. In fact, the first depot had been opened in the same neighborhood four years earlier by a New York pediatrician, Henry Koplik. Pure pasteurized milk was sold at Straus's depot for four cents a quart, and free milk was available for those who could not afford even this extremely low price.

Milk from cows inspected by veterinarians of the New York Board of Health was now being shipped to the city in ice-cold railroad cars. Straus had his own plant, where milk was kept on ice until it was pasteurized and bottled. Crowds came to the depot every day

Nathan Straus's milk station in City Hall Park, Manhattan.
(Museum of the City of New York/Art Resource, NY)

to buy fresh milk and sit in a tent by the river where doctors were on hand to dispense medical advice and examine children.

Next, Straus opened five more depots in other New York neighborhoods; in their first year, they dispensed 300,000 bottles of pasteurized milk. Soon he had twelve depots in New York City, all of which operated at a financial loss. In fact, the depots cost him more money than his share of the department stores' earnings. But he was on a mission. And he believed that milk was the perfect food, the perfect balance of protein, carbohydrates, and fat.

Since it was widely believed at the time that pasteurized milk had an odd taste, Straus also set up stands in the parks where the public could sample it for a penny a glass. He decided to launch a national campaign and wrote letters to the mayors of the largest cities offering to establish milk depots.

In his crusade for pasteurization, Straus often referred to an incident that had occurred on Randall's Island in the East River between Manhattan and Queens. The island was being used as an orphanage, and in order to ensure that the children had a ready supply of good, clean, fresh milk, a dairy herd was maintained there. But between 1895 and 1897, while the 3,900 children were being fed supposedly safe raw milk, 1,509 of them died.

In response to this frightening statistic, Straus built a pasteurization plant on the island. He made no attempt to change the children's diet or improve the orphanage's hygiene, just pasteurized the milk. The mortality rate declined from 42 percent of the children to 28 percent. A 28 percent child mortality rate would still horrify most people today, but in 1898 it was considered a remarkable improvement.

Milk depots were established in Boston, Chicago, Philadelphia, Cleveland, Chicago, and St. Louis. But many people in both Europe and America, and even in France, where Pasteur was revered, did not like pasteurized milk. The public complained about its taste. British farmers complained that the pasteurization machines were too expensive. Some doctors argued that milk lost its nutritional value when it was pasteurized.

In the United States, people also fought about a proposal to make pasteurization mandatory and raw milk illegal. In the spring of 1907, advocates of pasteurization, led by Straus, proposed an ordinance banning the sale of raw milk in New York City. At a rally for the ordinance, Straus said, "The reckless use of raw, unpasteurized milk is little short of a national crime . . ."

Dairy farmers who did not want to pay for pasteurization opposed the ordinance, as did those who thought that pasteurization would give farmers a false sense of security and lead to a decline in farm hygiene. Some even argued that it was better to eat live bacteria then dead ones—pasteurized milk being nothing more than milk with dead organisms floating in it.

The alternative was still certified milk, developed at Harvard Medical School in 1891. Rather than give in to cooked, dead, pasteurized milk, raw milk could be produced with greater care. Everything could be closely monitored—from the health of the herd to the hygiene of the farm to all the stages that milk passed through until sold in the market. But certified milk was expensive to produce and usually available only through doctors.

Advocates of pasteurization often acknowledged that raw milk had more nutrients than pasteurized milk, but they still insisted that pasteurized milk was safer to drink. Advocates of certified milk, or "clean raw milk," said that it, too, was safe, and continued to lobby for their cause. Among their supporters was the New York City Health Department, who was adamant that the solution lay in more rigorous inspection. Straus's proposed ordinance was defeated in May 1907.

What was being called the Milk Question attracted the interest of President Theodore Roosevelt, a native New Yorker with a reputation as a reformer. He ordered the Public Health Service to study the issue. They commissioned a panel of twenty supposed experts on dairy and in 1908 published a report concluding that raw milk was dangerous and that pasteurization did not alter the composition or flavor of milk. Many disagreed with the panel's findings, as is still the case today.

The pasteurization cause was not helped by the fact that other studies of boiled and evaporated milk showed that cases of rickets, a debilitating bone disease, could be traced to the "new milks." A number of cases of scurvy were traced to a lack of vitamin C in the milk.

In August 1908, Chicago became the first city to mandate the pasteurization of milk. After January 1909, all milk sold in Chicago had to be pasteurized. The only exception was raw milk from cows that had tested tuberculosis-free for a year. This was very close to saying that certified milk was also approved. The state's dairy farmers fiercely opposed the ordinance and went to court, claiming that it was a violation of free trade.

In New York in 1909, Straus tried again to have his proposed ordinance passed and was again turned down. But a year later, the city's health department changed its position and ruled that all

Union Settlement House, New York City, free milk program. Photo by Roy Perry. (Museum of the City of New York/Art Resources, NY)

milk for drinking had to be either boiled or pasteurized. In 1911 the National Commission on Milk Standards accepted both certified and pasteurized milk and declared that all milk should be one or the other. The American Medical Association came to a similar conclusion.

Meanwhile, Straus continued to argue that his ordinance would save babies' lives. He said, "Its defeat means babies killed." The ordinance came up for a vote again, and this time an overwhelming majority voted for it. In 1912 it became illegal to sell unpasteurized milk in New York. By 1914, 95 percent of New York City milk was pasteurized. By 1917, 46 major U.S. cities were requiring milk to be pasteurized.

At first, the method of choice for pasteurization was what was called "the flash method." Milk was heated to 184 degrees Fahrenheit for only a few seconds. But then it was found that more bacteria were killed by "the holding method," the one used at Straus's milk depots. It involved heating the milk to a lower temperature and holding it there for twenty minutes. This method became the established approach as the entire country moved to pasteurized milk.

Pasteurization was a public health decision, not a medical one. It is perfectly possible to ensure the safety of raw milk, as is sometimes done in other countries. But pasteurization is far easier to enforce; guaranteeing raw milk is a complicated and expensive process. This is of no comfort to people who prefer raw milk either for its taste or because they believe that it is more healthful.

Once Americans became confident that the milk they loved was safe, it took on what was at times an almost bizarre importance. In 1923 the future President Herbert Hoover, then Secretary of Commerce, addressed the World Dairy Congress with the words:

> Upon this industry, more than any other of the food
> industries, depends not alone the problem of public health,
> but there depend upon it the very growth and virility of the
> white race.

13

INDUSTRIAL COWS

THE ANGLO-SWISS CONDENSED Milk Company began operating in Europe a decade after Borden started up his company in America. By the 1880s, the European firm was producing 25 million cans of condensed milk a year. Condensing milk was a way of preserving it, which was a welcome idea in places such as Wisconsin and Switzerland, where milk was being produced faster than it could be sold.

Condensed milk also offered a solution for Australia, which depended on long-distance export markets. Toogoolawah, a small town in South Queensland, got its first condensed milk producer in the first decade of the twentieth century. In 1929, Nestlé merged with Anglo-Swiss, bought the plant, and moved it to Victoria as part of a plan to set up condensed milk plants throughout Asia.

Once milk was pasteurized and deemed safe, the principal reason for using condensed milk was that it was less expensive than fresh milk or cream. Sarah Rorer, in her 1913 book on ice cream and frozen puddings, said, "Ordinary fruit creams may be made with condensed milk at a cost of about fifteen cents a quart, which of course, is cheaper than ordinary milk and cream." In 1927, the General Electric Company published a book on cooking called *The Electric Refriger-ator Recipes*. In it was this advice on "How to Use Evaporated Milk":

When it is inconvenient to use heavy cream, evaporated milk can be used instead with excellent results. The mixture is not as rich or

1883 condensed milk advertisement for the Anglo-Swiss Condensed Milk Company, founded in 1881 by Charles Page, the American consul in Zurich. Seeing the success of Borden, he started his company with Swiss milk but aimed at the British market. The company became one of Borden's main competitors until it merged with Nestlé.

expensive as a frozen dessert made with cream. It will be smooth and not icy unless it stands in the refrigerator for an unusually long time.

Put evaporated milk in top of double boiler and heat over water. Let it cook 3 to 4 minutes or until milk is scalded. Pour into bowl. Cool in a room temperature pan. The flavor is more delicate if it remains in the refrigerator for several hours, without freezing. Beat with an egg beater until very light. Evaporated milk must be scalded and then chilled before it can be whipped like cream. One cup will increase in bulk two or three times. It may be used unbeaten if preferred. It can replace the cream in any mousse or ice cream which contains gelatin.

Fannie Farmer gave a clear explanation of how to scald milk in her 1896 book:

Put in top of double boiler, having water boil in under part. Cover and
let stand on top of range until milk around edge of double boiler has
a bead-like appearance.

Condensed milk became profitable, and a number of companies
started producing it. The quality of the cans improved and the milk
improved. In 1909 the problem of fat separation, reduced through
low-heat evaporation but never completely eliminated, was at last
resolved by a process called homogenization. It is one of those ideas
that seem obvious once someone thinks of it. The milk is forced
through a very fine screen so that the fat globules become tiny particles, and so no longer separate. The idea had originally been developed
in France in the 1890s by Paul Marix for making margarine. Then
other Frenchmen also working on artificial butter further refined
the idea. Once homogenization came into use, evaporated milk no
longer separated in the can, and its shelf life was greatly increased.

The Paris World's Fair of 1900 featured the debut of a new kind
of engine developed by Rudolf Diesel, talking films, and an esca-
lator. The new Art Nouveau design was also on display, as well as an
exhibit on the contributions of African American people to American society by W.E.B. Du Bois and Booker T. Washington. Less
noticed was Auguste Gaulin's *lait homogenisé*, the world's first homogenized fresh milk. It had been developed the year before in Paris by
Gaulin, the owner of the Gaulin Dairy Machinery Equipment
Company. Visitors at the fair were uncertain what to think of this
invention, which they called *lait fixé*, fixed milk, but after the fair
ended, Gaulin improved his machine and it became a huge success,
used in the making of evaporated milk and industrially made ice
cream.

But while the milk industry embraced homogenized milk, the
general public did not like their fresh milk homogenized. They
missed the "creamline," the place where the cream met the milk in
the bottle. And so the milk industry came up with a number of
strange and unappetizing demonstrations to convince the public

of the value of homogenization. One demonstration was conducted by the McDonald Dairy in Flint, Michigan, soon after they started selling homogenized milk in 1932. In an extraordinary statement, they informed the public that they had hired "professionals" to drink both regular and homogenized milk "under controlled laboratory conditions" and then to regurgitate the milks. This clearly showed that the curds from the homogenized group had been better digested. The vomited curds were then placed in jars in formaldehyde and given to milkmen show to customers along their routes.

When I was a boy, in the 1950s, a milkman delivered bottles of milk to our home in a metal carrying rack. The top part of each bottle, all the way down the neck, was a darker color than the rest of bottle—the cream had separated out. You shook the bottle and then poured. My brothers and sister and I loved that milk, and we each had a tall, cold glass of it every morning. But one day we didn't like the milk anymore. There was something wrong with it. It was all one color. We tried several different brands, but they were all the same. Milk wasn't as good anymore and we didn't love it the way we used to. The age of homogenized milk had begun. Soon nearly all milk in America was homogenized.

CONDENSED MILK BECAME a key ingredient in a number of sweets and caused a growth in the popularity of fudge in Britain and the United States. The origin of fudge is uncertain, but it is generally thought to be an American invention from the 1880s. By then condensed milk was commonly available, but it does not seem to have been used at first to make fudge. The first documented making of fudge was in Baltimore by a man who sold it for forty cents a pound. A Vassar student, Emelyn B. Hartridge, acquired his recipe, which used fresh milk:

2 cups granulated white sugar
1 cup cream

2 ounces unsweetened chocolate, chopped
1 tablespoon butter

Combine sugar and cream and cook over moderate heat. When this becomes very hot, add the chocolate. Stir constantly. Cook until mixture reaches soft ball stage (234°–238°F). Remove from heat and add butter. Cool slightly, then mix until fudge starts to thicken. Transfer to a buttered tin. Cut into diamond-shaped pieces before fudge hardens completely.

From Vassar, the making of fudge spread to sister schools Wellesley and Smith. None of the recipes used condensed milk, and fudge made with fresh milk is difficult to execute successfully. But by the turn of the century, fudge makers had discovered condensed milk, and in time condensed milk became a standard ingredient in the sweet. Later, marshmallow fluff became another essential ingredient for making a good fudge.

Marshmallow fluff was invented in 1917 by Archibald Query in Somerville, Massachusetts. He made and sold it out of his home. After World War I, two Massachusetts veterans, H. Allen Durkee and Fred L. Mower, commercialized it. Fresh from the battlefields of France, they called it Toot Sweet Marshmallow Fluff at first, but soon dropped the "Toot Sweet" because no one got it. By the 1930s they were sponsoring radio programs all over New England, and marshmallow fluff was becoming well known, though it is not certain who first used it in fudge. Here is the definitive fudge recipe from the doyenne of American cookie recipes, Maida Heatter. Note that "condensed" and "evaporated" milk are interchangeable terms:

Optional: 7 ounces (2 cups) pecans toasted or walnut halves or
    pieces
5 ounces (about 2/3 cup) evaporated milk
1 jar (7 ounces) marshmallow creme
1/2 stick (1/4 cup) unsalted butter

1 1/2 cups sugar

1/4 teaspoon salt

12 ounces (2 cups) semisweet chocolate morsels

1 teaspoon vanilla extract

Line an 8-inch square pan with aluminum foil: Turn pan upside down, center a 12-inch square of foil shiny side down over the pan, press down the sides and corners of the foil to shape it to the pan, remove the foil, turn the pan right side up, place the shaped foil in the pan and gently press the foil into place in the pan. Set aside the lined pan.

Pick over the optional nuts carefully (sometimes they include a piece of shell), and remove and reserve about ½ cup of the best-looking halves or pieces to decorate the fudge. Set nuts aside.

Pour evaporated milk into a heavy saucepan. Add the marshmallow creme, butter, sugar and salt. Place over low heat; stir constantly with a wooden spatula until the mixture comes to a boil. This mixture wants to burn; adjust the heat as necessary and scrape the bottom of the pan occasionally with a rubber spatula to be sure it is not burning.

As soon as the mixture comes to a full boil, start timing it; let it boil, stirring continuously, 5 minutes. (The mixture will caramelize slightly. It is not necessary to test the mixture with a thermometer—just time it—the temperature will be 226 to 228 degrees when the boiling time is up.)

Remove saucepan from heat; add morsels. Stir until melted and smooth; stir in vanilla. Stir in 1 1/2 cups of the nuts. Quickly pour into the pan; smooth the top. Place remaining ½ cup nuts on top, spacing them evenly and pressing down enough so they will not fall off.

Let stand until cool; chill until firm. Remove the fudge from the pan by lifting the corners of the foil. Carefully cut the fudge into pieces. Wrap individually in clear cellophane, wax paper or aluminum foil. Or place the fudge in an airtight freezer box. If you want to store for more than a few days, freeze it.

Note: To toast nuts, place in a shallow pan in the middle of a 350 degree oven, stirring occasionally, until very hot but not until they become dark, 12–15 minutes.

In British-ruled late-nineteenth-century India, enormous amounts of condensed milk were used to feed babies. The Indians also started using sweetened condensed milk for their tea and even for their frozen dessert, *kulfi*, altering a recipe dating back to the sixteenth century. In China, where condensed milk was also imported for babies in the nineteenth century, it started to be used in one of their rare milk dishes, "fried milk."

In the Dominican Republic, *batida de lechosa*, virtually the national refreshment, is made from condensed milk, papaya, and sugar put through a blender, sometimes with a few drops of vanilla extract added.

Entirely new dishes were also created. In Argentina, *dulce de leche*, a thick dark caramel sauce made with reduced sweetened condensed milk, became a national dish. There are numerous ways of cooking down condensed milk, but the popular technique is to just put the can in a pot of boiling water and leave it there for four or five hours. Simple, but there is a catch. If the water is allowed to boil away to the point where part of the can is exposed to air, it will explode.

AN ODDITY OF the milk business in America and in Europe was that its growth was not determined by demand. Sometimes production grew faster than demand. What was motivating farmers was that the price for milk was so low that a farm had to have more cows and produce more milk just to stay viable. In the nineteenth century, the New England family farm with only three to five cows found survival nearly impossible. The trend continued and by the twentieth century, the forty-cow farm was finding survival difficult.

As Americans moved west, to places with more open land, the number of cows in America kept growing. In 1850 there were

6,585,094; in 1900 there were 17,135,094, almost triple the 1850 number.

The importance of breeding was recognized in Europe long before it was recognized in America. The English farmer William Ellis wrote in 1750, "She was a healthy one, hardy, gentle, and easy milked. Such a cow as this deserves to have her breed increased." In the eighteenth century, British farmers started realizing that the quality of the bull had much to do with the quality of the cows he sired. In 1726, John Lawrence, "a country gentleman," penned a book called *New System of Agriculture*. In it he wrote that more attention needed to be paid to the bull because "the males of all creatures are the principal in the breed and generation." Though his logic seems a bit dubious, his fundamental insight, that it took more than a good cow to give birth to another good cow, was an important one. British farmers began to realize that it was a waste to have a high-quality cow breed with any available bull, or as they put it, "everybody's son." A list of desirable bull traits started to be established: A bull should have a broad forehead, his eyes should be large and black, his horns should be long, his hair smooth as velvet, his neck thick, his chest big, his buttocks square, and more. Different farmers had different standards, and different breeds were created. Owing to the reputation of Dutch cows to be productive milkers, they were brought in to English farms to either serve as a separate breed or be used for crossbreeding.

Originally most European livestock was bred for the quality of its beef or its strength as draft animals. But in the eighteenth century, some of the new breeds, such as the Jerseys and Guernseys from the Channel Islands, started to earn reputations for the quality of their milk. The Ayrshire breed in southwestern Scotland, an early breed from the seventeenth century, was not developed for dairy, but in the eighteenth century, it was realized that these cows produced an unusually high yield of milk that was also particularly good-tasting. It had a high lactose content, giving it a sweet taste. It also had

Ayrshire

unusually small and uniform fat globules, which before homogenization was a highly desirable trait.

In the second half of the nineteenth century, the best cow breeds were brought from Europe to America. Among them were the British Shorthorns, which for a time became the leading American dairy cow; Holstein-Frisians from North Holland; Ayrshires from Scotland; Jerseys and Guernseys and Red Polls from the south of England; Brown Swiss and Simental from Switzerland; and Normandys from France. One of the most appreciated British breeds of the nineteenth century was the Alderney, a small cow of Norman origin from the Channel Islands, cited by Mrs. Beeton as having the richest milk. A few Alderneys were imported to America, after which the pure breed became mixed out of existence in the Channel Islands. According to some cattle experts, the closest thing to a true Alderney today is found in America. But this claim can also be disputed.

The United States now had the best breeds from all the best milk regions. Improved cow breeds pushed the average annual yield of a cow from 1,436 pounds of milk in 1850 to 3,646 in 1900. America was awash in milk. Fortunately, there were products besides fresh milk that had room for industrial growth—condensed milk, baby formula, and cheese were all growing markets.

Brown Swiss

The distance that a dairy product could travel started to increase in the 1840s with the development of the railroads. This element added to the farmers' cost of production but also created new opportunities for cheese. Until the 1850s, farms did not produce enough milk to support a cheese factory. Each farm might make a few cheeses. In New England they formed cooperatives, with farmers pooling their milk for cheese production, and such cooperatives are still common there.

The first permanent cheese factory and industrial-scale cheese-maker in America was established in Rome, New York, in 1851 by Jesse Williams, who came from a cheesemaking family. Having learned the lessons of the Industrial Revolution, Williams produced cheese on an assembly line. A decade later, factory cheese became easier to make when mass-produced rennet became available.

For years there were many who thought a factory system wasn't suitable for cheesemaking. From a gastronomic perspective, that is still a fair subject for debate, but by the end of the century there was little doubt that cheese factories were economically viable. This was the age of factories, and they were the heralded solution to most every problem.

Xerxes Addison Willard, a leading spokesman for the dairy industry, wrote in 1865:

The questions have been frequently asked: Is the factory system destined to stand the test of years? Is it to continue to prosper? Or will it soon break up and dairymen return again to the old order of cheese making? In my opinion, it is to live. The system is a progressive step, and all history teaches that when that is taken it is difficult to retrace it.

Cheese factories did prosper, but artisanal cheesemakers have also survived.

Wisconsin had almost too much milk, and the state started becoming a large-scale cheese producer. Who first industrialized Wisconsin cheese is not clear. It is often said that the first cheese factory in Wisconsin was built by Hiram Smith in Sheboygan County in 1858, but he stopped making cheese after a year and the factory became a fresh milk plant. In 1864, Chester Hazen built a cheese factory in Ladoga, also often cited as the first Wisconsin cheese factory.

In the mid-1870s, John Jossi of Dodge County, Wisconsin, started producing his own type of cheese, which he called brick and made in a factory owned by the Swiss Cheese Company in Wisconsin. He then opened other factories around Wisconsin. His company, in continuous production until the end of 1943, was eventually acquired by the Kraft Cheese Company.

By the turn of the century, Wisconsin was the leading cheese-producing state, with fifteen hundred factories of varying sizes.

The Europeans were also starting to build cheese factories. In fact, the world's first cheese factory was built in Switzerland in 1815, but it was not commercially successful. In England, the first cheese factory was opened in 1871, and in the Netherlands, in the 1880s. In the twentieth century, artisanal cheese went into rapid decline.

In 1889, Adolph Tode, owner of the Manhattan Delicatessen in New York City and the Monroe Cheese company upstate, was having difficulty finding a reliable supply of a popular German cheese, Bismarck Schlosskäse. He asked his cheese company to produce it locally. A twenty-two-year-old, Emil Frey, started to work on it and in 1892 produced a cheese that, although not exactly like the German cheese, was close to it. In fact, Tode and others thought it was better. They named it Liederkranz; *liederkranz* means "singing society" in German, and Frey was a member of a singing group of that name. At first, Liederkranz cheese was a uniquely New York City food, but word of it spread and by 1926 there was not enough milk in Monroe County, New York, to make it. Production had to move to Van Wert, Ohio, in the Midwest, where there was an ample milk supply.

In 1918, Frey invented Velveeta, which incorporated whey, unlike other cheeses where the whey was pressed out of the curd. It was marketed as a velvet-textured cheese—hence the name Velveeta—and touted for its extreme meltability. It was so successful that a separate Velveeta cheese company was formed in 1923. It, too, was acquired by Kraft, in 1927.

WITH AN ABUNDANT availability of factory-made American cheeses that melted well, America developed a melted-cheese cuisine—cheeseburgers, grilled cheese, macaroni and cheese—that became quintessentially American. But melted-cheese cooking was not new, nor was it uniquely American. In Italy there was pizza, and in Brittany, cheese crepes. The Swiss national dish, dating from the sixteenth century, is melted cheese and wine in the form of fondue. Americans started to embrace them all.

There was also Welsh rabbit, which in the twentieth century became popular in a number of countries, including the United States. Today the melted cheese dish is usually called "rarebit," meaning a tasty morsel, because it so obviously isn't a rabbit. It is assumed that "rabbit" was an erroneous label. The earliest record in 1725 called it rabbit and no one used "rarebit" until 1780. It is also not clear if it is originally Welsh. The English like to use the term "Welsh" pejoratively—when something is second-rate or fake, they

Greet "the gang" with this good-for-'em cheese food

IT'S DIGESTIBLE AS MILK ITSELF

You can be indulgent about between-meal snacks when you have Velveeta in the house. Because, *like milk*, this top-quality cheese food helps supply the children with important food values they need: milk protein, milk minerals, vitamins A and G. And delicious, golden Velveeta is as digestible as milk itself!

That's why wise mothers insist on Velveeta. In nutrition and in flavor it's the finest of pasteurized process cheese foods. Always keep Velveeta on hand to satisfy young appetites at any hour, or to use for meal-time sandwiches too.

For a backyard picnic: Velveeta-Stuffed "Franks"

Slice frankfurters lengthwise and spread generously with golden Velveeta—the *quality* cheese food that's rich in important food values from milk. Wrap a slice of bacon around each frankfurter and fasten with toothpicks. Place under low broiler heat until the Velveeta is melted and the bacon crisp. Serve on split, toasted frankfurter rolls.

VELVEETA

THE QUALITY CHEESE FOOD ...MADE BY KRAFT

in ½-lb., 1-lb. and 2-lb. pkgs.

1960 advertisement for Velveeta cheese.

call it Welsh. So calling the dish a Welsh rabbit might have been a bad English joke. Was the dish originally English? The cheese traditionally used to make it is either Gloucester or cheddar, which are both English cheeses.

But the Welsh are also famously passionate about melted cheese. In the fourteenth century a humorous tale was written in which the Welsh were all loitering in heaven. To get rid of them, Peter stood outside the gates and shouted *"Caws pobi!"* which means "toasted cheese" in Welsh. The Welsh all ran out to get some, and Peter slammed the gates closed.

The Scots are equally fond of toasted cheese, as recalled in Robert Louis Stevenson's 1883 novel, *Treasure Island*. The wretched castaway Ben Gunn longs for cheese, saying, "Many's the long night I've dreamed of cheese—toasted, mostly."

Hannah Glasse offered recipes for Scotch, Welsh, and English rabbit. This is her Scotch rabbit:

Toast a piece of bread very nicely on both sides. Butter it. Cut a slice of cheese, about as big as the bread, toast it on both sides, and lay it on the bread.

Glasse's Welsh rabbit was almost identical but added mustard. The English added wine. Today the dish is usually made with beer.

The French love Welsh rabbit, which they always call *un Welsh*. They love it because even more than the Americans, the Scots, the Welsh, and the Swiss, the French love melted cheese. French literature is full of melted cheese. Alphonse Daudet included a story in his *Contes du Lundi* about the impact on a room of the smell of cheese soup—*"Oh! La bonne odeur de soupe au fromage."*

The Americans could not have eaten all these melted cheese dishes until the Industrial Revolution. Only after they had an abundance of factory-made cheese did the dishes become widespread. Here is a cheese soup made by Chef Stanley Hamilton at his restaurant in

Union Station, St. Louis. President Harry Truman frequently dined there, and this soup was said to be a favorite of his. Hamilton called the recipe "Cream of Wisconsin Cheese Soup":

12 saltine crackers
1 qt beef broth
3 cups grated sharp cheddar cheese
3 tbsp butter
3 tbsp all-purpose flour
1 tbsp Worcestershire sauce
1 cup light cream
¼ tsp white pepper

Place saltines in oven to warm. In a saucepan cook two cups of broth over medium heat. Add cheese, stirring constantly as it melts. Add remaining broth and simmer until smooth. Meanwhile, in a small skillet over medium heat, make a roux with butter and flour. When smooth add to first mixture. Continue stirring as you slowly add cream, Worcestershire, and pepper. Stir constantly at simmer for fifteen minutes. Serve with toasted crackers.

Even before industrialization, the French produced enough cheese—much of it of the Gruyère or Swiss type—to indulge in their passion for cooking with it. Whether it was onion soup or fish or scallops in a savory tart, there was always the possibility of presenting the dish *au gratin*, that is, with melted grated Gruyère on top.

Claude Terrail, the tall and elegant owner of Paris's La Tour d'Argent restaurant, was one of that melted-cheese-loving generation of Frenchmen. His old-fashioned cuisine, served in a restaurant with one of the best nighttime views of Paris, brought him three Michelin stars in the 1950s. In time including an abundance of melted cheese seemed old-fashioned, and his rating was downgraded.

Best known for his pressed duck, Terrail also made dishes such as filet of sole topped with béchamel sauce and melted cheese. He loved Welsh rabbit and insisted that it be made with Gloucester cheese:

You begin with 250 grams of this cheese cut in small cubes or grated, a half-teaspoon of English mustard, and a deciliter of English ale. Put on low heat and stir with a fork: when heated the cheese is liquefied and it is not difficult to blend it with the beer.

Once perfectly blended, you can pour it over English bread, sliced and buttered.

Broil under high heat.

In 1848, Ferdinand Hédiard returned to Paris from the tropics, bringing with him such exotic fruits as mangoes, bananas, and pineapples. He then opened a store on the Place de la Madeleine, where Parisians have been shopping for *exotique* products ever since, despite the fact that many of these wonders are now commonplace. Hédiard and his wife and daughter also created recipes to show Parisians how they might use these strange and interesting fruits. And of course one way was to serve them *with melted cheese*. Here is his recipe for *gratin de bananes*:

Cut off the ends of plantains and cut them lengthwise and boil them for five minutes in their skins. Once they are cooled peel them and sauté them in hot butter and with that butter make a light béchamelle with 1 liter of milk seasoned with cinnamon; add a pinch of cayenne pepper. Cut the plantains into discs and place them in a buttered mold or gratin dish. Pour béchamelle over and then sprinkle with grated gruyère cheese. Put it in the oven for five minutes and serve hot.

ONE OF THE impacts of industrializing dairies and cheesemaking was that it took jobs away from women. On farms around the world,

Milk below Maids *by Luigi Schiavonetti (1765–1810), milk vendors on a London street.* (HIP/Art Resource, NY)

women had traditionally been responsible for milking the cows and making the cheese and the butter. The work was hard and the hours were long, but the dairymaid was romanticized, as in an 1830 American song about Fanny the milkmaid entitled "Pretty Maidens Here I Am."

In cities, most milk vendors were also women—hardy women who went out on the streets with seventy-pound buckets and worked long days. In London they famously cried out "Miow," which meant "Milk below." Even on more affluent farms with employees, women were hired for the dairy-related tasks, and they answered to the lady of the house.

The dairymaid was always a selling point for milk products. She was an idealized, vaguely sexual image of wholesome femininity.

M for Milkmaid *by William Nicholson, from an*
alphabet in lithographs from an original series of
woodcuts, 1898. (Author's collection)

In the mid-seventeenth century, Izaak Walton, who aspired to the literary heights of his friend John Donne, in his book on flyfishing, *The Compleat Angler*, apropos of absolutely nothing, midbook has an entire chapter of various poems and discussions he wrote on the wonderfulness of milkmaids. He spoofs his contemporaries— Christopher Marlowe, Walter Raleigh, and John Donne—suggesting their love poems were about milkmaids, and makes the curious observation: "our good Queen Elizabeth did so often wish herself a Milkmaid all the month of May, because they are not troubled with fears and cares, but sing sweetly all the day." In the romantic version of milkmaids there is never mention of how hard they worked.

In 1784 Josiah Twamley, a writer concerned with quality and efficiency in English dairies, wrote:

Nothing is more commendable in a dairy-maid than cleanliness, nor will anything cause them to be more esteemed; everyone who perceives extreme neatness in a dairy, can not help wishing to purchase butter or cheese from so clean and neat a place.

But with industrialization, women were gradually pushed out of dairies, except on small family farms. The trend had begun even earlier, in the eighteenth century. Cheddar was a forerunner of industrial cheese because of an early-nineteenth-century cheddar maker, Joseph Hardy, who wanted to produce as much cheese as possible as efficiently as possible. Many of his ideas led to the cheese factory, and he was one of the first to employ male cheesemakers. Part of their large-scale production was to make huge individual cheeses that were thought to be too big and heavy for women to handle.

It was then, too, that some dairies began switching to male managers, something new at the time. Women could do farm work, but industry was for men. There was a widely held prejudice against women in factories. In some countries there were laws against women working in factories at all, or working on Sundays, or working night shifts. Initially, many cheese factory owners objected to these new laws. In Holland, cheesemakers said that cheese could not be made without the expertise of women. Gradually, though, they learned that cheese could be made effectively with an all-male staff.

THE CENTRAL TECHNOLOGY that allowed the dairy industry to become truly an industry was machine milking. Hand milking is a slow and difficult process. The reason why most pre-nineteenth-century herds seldom grew to more than forty cows was that a family would not have enough time to milk them unless they hired a large staff, which would cut into their already narrow profit margin.

In the nineteenth century, Catherine Beecher gave a good description of milking, a process that had not changed since ancient times:

> In milking, put the fingers around the teat close to the bag; then firmly close the forefingers of each hand alternating, immediately squeezing with the other fingers. The forefingers prevent the milk flowing back into the bag, while the others press it out. Sit with the left knee close to the right hind leg of the cow, the head pressed against her flank the left hand always ready to ward off a blow from her feet, which the gentlest of cows may give almost without knowing it, if her tender teats be cut by long nails, or if a wart be hurt or her bag be tender. She must be stripped dry every time she is milked, or she will dry up; and if she gives much milk it pays to milk three times a day, as nearly eight hours apart as possible. Never stop while milking till done, as this will cause the cow to stop giving milk.

Why would this description be included in a book called *The American Woman's Home*? Perhaps sometimes small-scale farmers with no dairy experience decided to take on a cow or two.

As complete as Beecher's description seems, she does not include a few things, such as how much your hands and forearms ache when you cannot let up, or how a grumpy or playful cow can hurt you. Cows are very large. And if you have several cows and only eight hours between milkings, there is little time between finishing the last cow and remilking the first one.

In the nineteenth century, when a new machine seemed to appear every month, many people began trying to develop a machine that could milk cows. But how could a machine accomplish this sensitive and complex task? The closer the inventors came to achieving this goal, the more it was questioned. In 1892, S. M. Babcock, famous for inventing a very usable device for measuring the fat content of milk, wrote in the *National Dairyman* that "milking machines would result in poorer quality of milk and lowering the standards of dairy animals."

The first milking machine idea, developed midcentury, was crude—insert a tube up the cow's teat, forcing open the muscle and allowing the milk to flow down the tube into a bucket. The original tubes were wood, but better ones made of silver, bone, or ivory were later used, and some were still selling in the early twentieth century. The British developed the first patent on tube milking in 1836. But tube milking spread disease in both cows and humans, and it often injured cow teats so badly that some permanently leaked milk.

Then the British became interested in the use of pumps. Between 1860 and 1862, a number of inventors took out patents on machines with cups for the teats and a pumping suction. In 1889 Scotsman William Murchland built a well-functioning vacuum pump milker that hung under the cow. Many other machines with hand pumps were also invented.

In 1898 much attention was paid to William Mehring's milker because it could milk two cows at once. After all, the entire object was to milk more cows faster. The Mehring milker was operated by using foot pedals to create a vacuum, and it was still in use in the early decades of the twentieth century.

It is odd that in the age of steam power, no one thought of inventing a steam-powered milker until 1898. (The first steam-powered tractor was invented in 1868.) Called the Thistle Milker, it was designed by Alexander Shields of Glasgow and featured a steam-powered vacuum. But like all early mechanical milkers, it had a serious drawback—it could not compensate for the changing size of the cow's teat, which becomes smaller as the milking progresses. The mechanical milkers also pumped some of the milk back into the udder.

In the 1890s the problem was solved by a number of inventions including a machine by Alexander Shields that massaged fluid from the teats and a double-chambered teatcup, which was also a teat-massaging device. Now farms could have far larger herds without maintaining a large staff of workers. Twenty or forty or more cows could be herded into a milking parlor. In the front of each milking stall was a tray filled with forage for the cows to nibble on so that they would enjoy their milking. Four cups were placed on the four teats of each cow. The cups were removed—with today's modern machines, they fall off automatically when done. The cows were herded out and a new group herded in. The parlor could be operated around the clock with some time off for cleaning. This serviced a lot more cows than ever dreamed of by Mrs. Beecher.

Certified milk producers were still interested in showing that their dairying ways were the best, and at the 1939 New York World's Fair, the Walker-Gordon Certified Milk Dairy, a New Jersey certified milk company, displayed a new milker called the Rotolactor, quickly dubbed "the cow merry-go-round." The milking parlor was a revolving disk with fifty stalls around it. Cows lined up and were led one by one into a stall. A milker washed her udder with a clean cloth and attached four cups to her teats. The milking machine was turned on and the cow rode around the disc for ten minutes until her stall arrived back at its starting point, where the cow's cups were detached and she was led off. In this manner the merry-go-round milked three hundred cows every hour.

At the World's Fair, the new invention was proclaimed to be the milking machine of the future. Walker-Gordon's had wanted to demonstrate the high standards of cleanliness used in the production of certified milk. They did not care about manufacturing their display machine or promoting it as the milker of the future. Yet today, one of the most expensive and sought-after high-tech milking machines is the rotary milking parlor, built on very much the same idea as the Rotolactor.

★   ★   ★

STARTING IN THE mid-nineteenth century, the dairy industry became focused on the quantity of milk being produced. Given the very small profit margins of the industry, quantity was essential. And in the twentieth century, too, everything dairy related, including farm bulletins, agricultural college research, and advice to farmers, emphasized the importance of becoming larger—of investing in a larger herd of cows.

The smaller dairy farms were failing. In the nineteenth century, a larger dairy had meant forty cows, but in the twentieth century, it meant a hundred cows and then several hundred cows. By the end of the twentieth century in America, the world's largest milk-producing country, a large farm meant thousands of cows.

Milk prices had always been low relative to the cost of production. That has always been one of the incentives in cheesemaking. During the Great Depression, the public could not afford milk even at low prices and demand declined. Then prices sank and angry farmers across the country went on strike. The federal government stepped in to stabilize milk prices and that system has remained. The government establishes a milk price. It is usually affordable for large farms, but for small ones, it sometimes falls below their cost of production. They can charge more, but that milk would not be competitive with the milk priced at government levels. And once the interstate highway system was built in the 1960s, milk did not necessarily have to be locally produced.

Before World War II, 80 percent of milk in America was hand-delivered to doorsteps. Not until the second half of the twentieth century were deliverymen replaced by stores. Giant supermarket chains that worked with giant dairies and sold cheap milk took over. Cartons replaced bottles. They could be thrown away and didn't have to be washed and they didn't break. The milk inside them was no longer visible, but since homogenization, there was nothing to see anyway.

In the twentieth century, cows were expensive to buy and expensive to maintain. The number of cows per farm in the United States

"The milk will get through." This image of a milkman delivering in the London Blitz, 1940, is staged, with the photographer's assistant dressed as a milkman.
(HIP/Art Resource, NY)

has been increasing, but the total number of cows in the country has been declining because many of the smaller farms have closed. At the same time, the amount of milk produced has been dramatically increasing. According to the U.S. government, in 1944 the country had 25.6 million dairy cows. By the twenty-first century there were only 9 million, but those 9 million produce far more milk than did the 25.6 million in 1944. In 1942 the average cow produced less than 5,000 pounds of milk in her lifetime. Today that average is up to 21,000 pounds. At the same time, milk consumption, even with a far larger human population, has declined. This was not expected. In the nineteenth and early twentieth centuries, milk consumption rose fairly steadily.

Part of the increase in milk production per cow has been the result of higher-protein feed. Farmers today spend much more time growing high-protein crops such as alfalfa and corn than they did before. Today, even grass-fed cows are given fodder to get through the winter. Depending on what can be grown or reasonably purchased, a wide variety of crops—including oats, barley, corn, alfalfa, sorghum, millet, and clover—are stored for the winter. Some farmers ferment the mixture, which is called silage, or blend the cereals together, and these can be richer in protein than the grass that is available in the fields. This has created a long-standing argument about which way is the better way to feed cows.

Farmers are also spending more money buying feed concentrates. In 1900 there were only six thousand acres in the entire United States planted with alfalfa. By 1986, that number had risen to almost 27 million, and it is considerably more today.

Another change is the prevalence of breeding. High-producing cows and the bulls that sire such cows are now regarded as breeding stock. In the 1930s and 1940s artificial insemination was developed, which made breeding far more efficient. Today, a dairy farmer never has to see a bull, which makes most of them happy, because bulls are very difficult and dangerous animals. Farmers used to seek out a good-looking bull in the neighborhood. Today, the bull sperm can come from anywhere and is sold by performance records, not the appearance of the animal. Some argue that this is taking farmers further away from the true nature of their animals.

The cow that has been settled on as the best by most farmers all over the world is the Dutch Holstein-Frisian, sometimes just called a Holstein or a Frisian. They are actually one of the oldest breeds in America—the breed that the Dutch brought in 1613. Today these very large black-and-white cows are bred to produce large quantities of milk, a goal that can only be achieved if they are fed huge amounts of high-protein food. A group of scientists from the U.S. Department of Agriculture and the University of Minnesota have calculated that 22 percent of the Holstein genome has been altered by human selection over the last forty years.

Artificial insemination has made breeding extremely efficient. Sperm is produced in a limited number of places, and meticulous performance records are kept. Nothing like this could ever have been accomplished with naturally bred animals. Today's breeding centers can develop superstar bulls and use their sperm extensively. In May 2012, the *Atlantic* reported on a Holstein bull named Badger-Bluff Fanny Freddie, who was born in 2004. It was deemed ideal and had already produced 346 daughters.

But the formula for success is complicated. A cow has to be pregnant in order to lactate, so fertility is important, but it seems that the higher the cow's milk production, the lower its fertility.

Holstein

The Holstein-Frisian also has some disadvantages. They are extremely expensive animals to work with, as they consume an enormous amount of food.

Technically speaking, the Holstein-Frisian is not the world's most productive cow. There is a small cow from the island of Socotra in the Arabian Sea, part of Yemen, that produces more milk for the amount it is fed than any other cow in the world. It might seem as if this should be the most sought-after cow, but dairies generally go for the largest milk producer, no matter how much it has to be fed.

In America, milk advertising often shows a pretty brown Jersey— and they are pretty cows. But they are rare in America now. Big black-and-white Holsteins are everywhere, making up about 90 percent of the dairy cows in America. In Britain, too, it is rare to find a Jersey, Guernsey, Ayrshire, or Shorthorn. Black-and-white Holsteins are part of the landscape. It is the same in most of the rest of Europe, and in Asia and Australia as well. A few farmers, especially those who make cheese, still maintain traditional breeds that produce a higher-quality milk, but they are unusual. In fairness, the Holstein does yield a good-quality milk; it's just that the milk from a Jersey or Ayrshire is better. And there are others, but they are disappearing. The United Nations Food and Agriculture Organization estimates that two breeds of domestic animals become extinct every month.

Modern dairying has become a very tough business.

14

MODERN CUISINES

NINETEENTH-CENTURY AMERICANS, FRENCH, British, and Italians loved milk. It was an age of dairy sauces—milk, cream, and butter—milk and cream dishes, milk and cream soups, and milk and cream drinks.

The idea of cream sauces seems to have started in France. When the English started making milk sauces, they termed their dishes *à la crème*.

Savory cream sauces became one of the hallmarks of what was first known as "the new French cuisine" and later known as classical French cuisine. The taboo on mixing fish and dairy had long been forgotten, and dairy sauces were especially popular with fish.

The most famous fish-and-cream dish was *sole normande*. French food historians date this dish to an 1838 banquet, and it was probably the dish that made *normande* an adjective meaning "in cream sauce." Normandy was famous for its cream.

But classical French cooking was extremely rich and loved cream. Auguste Escoffier, born in 1846, was the French chef who carried nineteenth-century cooking into the twentieth century. His cooking and such books as *Ma Cuisine* and *Le Guide Culinaire* are said to define French classical cooking. Escoffier's motto was "Make it simple," though he almost never did. In classical French cuisine, sauces are made from sauces that are added to sauces, so they are usually made in restaurants rather than homes. All you need for Escoffier's *sauce normande*, though, is a good fish stock, another stock made from the trimmings of the sole after fileting known as *fumet de sole,* and a *cuisson de champignon,* which is minced mushrooms sautéed in butter, mussel broth, and a few other things. His *sauce normande* is intended to be served on top but is only one of the elements in his *sole normande.* Others include poaching the sole in fish stock and adding mussels, slices of black truffles, croutons, and shrimp. If you have the patience, time, and resources, this kind of cooking will reward you with a tremendous depth of flavor. Here is the sauce:

> Add to three quarts of fish stock, a deciliter of cuisson de champignon and broth from mussels; 2 deciliters of fumet de sole; several drops of lemon juice mixed with 5 egg yolks mixed with 2 deciliters of cream. Reduce by about a third on a high flame.
>
> Pass the sauce through a cheesecloth and then finish it with a deciliter of double cream and 125 grams of butter.

Escoffier said that while this sauce was intended for *sole normande,* it could also be used for many other dishes.

There are many easier ways to make a cream sauce. This is a recipe from Sarah Josepha Hale, the editor *of Godey's Lady's Book,* that appeared in her 1841 book, *The Good Housekeeper.* She recommended it for fowl, but it would also work for fish:

> Melt in a teacupful of milk a large table-spoonful of butter, kneaded in a little flour; beat up the yolk of an egg with a spoonful of cream, stir into the butter, and put it over the fire, stirring it constantly. Chopped parsley may be added.

Probably neither of these sauces would be made today by leading chefs, the Escoffier because it seems overly complicated and the Hale because the use of flour in sauces is out of fashion, except in New Orleans where traditions seldom die. But using flour is a very old idea. The Romans, including the celebrated first-century A.D. cook Apicius, made sauces thickened with either plain flour or *tracta.*

When Europeans started making savory cream sauces for their meat and fish dishes, they started by making a thickener of flour and melted butter. This technique later got a bad reputation because of heavy-handed cooks. The thickener, known as a *roux*, requires a light touch so that you don't end up with flour paste.

It is not certain whether the Europeans learned how to make roux from the Romans or if they simply reinvented it. There were numerous European names for the sauce of roux-thickened milk, but eventually it became known as *béchamel*.

François Pierre de la Varenne, popularly known as La Varenne, was the most influential chef of seventeenth-century France at a time when the new French cuisine was starting to move away from the lean, sharp, spicy, and acidic tastes of the Middle Ages and from Italian influence. La Varenne started several sauces with small amounts of flour and melted butter, but because he used a much larger amount of other ingredients, they did not become true milk or cream sauces.

Vincent La Chapelle, an eighteenth-century chef for French royalty, most famously Louis XV's mistress, Madame de Pompadour, proclaimed in his 1733 book *The Modern Cook* that French cooking had completely broken from its past. He offered a recipe called *turbot à la béchamelle*, thought to be named after his influential contemporary Louis de Béchameil, though if true, he misspelled his name. The turbot was poached in a *court bouillon*, literally, a "short broth," a quickly cooked stock. Minced parsley, green onions, and shallots were tossed in a pot with large pat of butter; salt, pepper, and nutmeg were added, followed by a small amount of flour. The fish was added next, and then La Chapelle offered a choice—the fish could be cooked with milk or cream "or equally well some water." Dairy had appeared in a savory dish, but it was still an afterthought.

Soon after that, however, the so-called modern French cuisine came to mean the use of milk and cream sauces. William Verrall, an English innkeeper in Lewes, Sussex, who had trained under a French chef, became known for introducing many new food ideas, including the French use of dairy sauces, to England. The family inn, the White Hart, dated back to the Elizabethan period, but under the Verralls it became fashionable. His 1759 book, *A Complete System of Cookery*, was intended, the author wrote, "to show, both to the experienced and inexperienced in the business, the whole and simple art of the most modern and best French cookery." The slim book was full of savory cream dishes such as gizzards and cream, peas and cream, spinach and cream, and "Breasts of fowls à la Binjamele." The spelling was getting ever farther away from the late Monsieur Bechameil's name, but the use of cream had become firmly established. Many of Verrall's recipes were unreadable, but his book was very influential nonetheless. Here is one of his poultry dishes. It includes an orange, which may have been his own touch. No longer grown solely in the orangeries of the wealthy, oranges were extremely popular in the eighteenth century:

> Two fowls make two dishes, but in different ways; cut off the legs whole with the feet, and the next shall give directions how to manage them. But the breasts you must roast, but without the pinions [the last joint at the end of the wing], they may serve for something else: when roasted, take off the skin, and cut off the white flesh, slice it in thickish pieces, put it into a stewpan, and provide your sauce as follows, take about half pint of cream, a bit of butter mixt with flour, put in a green onion or two whole, a little parsley, pepper and salt, stir it over a slowish fire till it boils to its thickness, pass it through an etamine [cheesecloth]. Put it to your fowl in a stewpan, and then boil it till it is hot through: add nothing more than the juice of an orange, and send it up.
>
> This sauce may serve for any sort of white meat, and is now very much in fashion.

Marie-Antoine Carême, a chef who straddled the eighteenth and nineteenth centuries, established the so-called *grand cuisine* that Escoffier inherited in the next century. He said that there were four kinds of base sauces, one of which was béchamel. Carême's béchamel as described in his 1817 *L'Art de la Cuisine*, was a daylong project. First, a light meat stock had to be made and used to cook a "quintessence" of meat by simmering various cuts of veal, ham, and fowl in it. After the stock was reduced, the meat had to be pricked to open it up and more stock added. Then it was cooked down again. Eventually a flour roux was added and everything simmered for hours with herbs. Next it was clarified, and finally cream was added. At the end of his recipe, Carême commented:

> It is essential that the reduction of this sauce be done by a man who loves his station: for an insouciant man will let it take colour, and that is the worst thing that could happen to this sauce, which, moreover, honors the Marquis de Béchamel, who thought of the addition of cream to velouté and gave his name to this excellent sauce.

Perhaps the reason Escoffier claimed his role as a simplifier, despite his complicated formulas, is that he came after Carême. His béchamel used only veal stock, and he claimed—some cried disgrace—that it could be made in an hour. Essentially the sauce is a butter and flour roux in a reduction of veal stock with milk or cream.

Cream sauces had also become popular in Italy. The Italian chef Pellegrino Artusi said that his béchamel sauce was like that of the French "except that theirs is more complicated."

In 1928 Ada Boni, the editor of the leading women's magazine in Italy, *Preziosa*, taking on the impossible task of writing the definitive cookbook of Italian cuisine, published *Il Talismano della Felicità*. Still regularly reprinted, it was the classic Italian cookbook, and in it were many cream sauces. Many started with flour and butter, but not all, including this unabashedly rich recipe for pheasant with cream:

1 small pheasant
3 tablespoons butter

½ onion, chopped
salt and pepper
1 ¼ cup heavy cream
1 tablespoon lemon juice

Place pheasant in Dutch oven with butter, onion, salt and pepper. Cook slowly, browning gently on all sides 2 ¼ hours. Add cream and continue cooking 30 minutes. Just before serving, add lemon juice to gravy and mix. Serve immediately.

Like Boni's pheasant, the most famous seafood-and-cream dish of nineteenth-century America, lobster Newburg, did not use flour. Louis Fauchère, a chef for New York's famous Delmonico's restaurant, opened a hotel restaurant in Milford, Pennsylvania, in which he claimed to have invented lobster Newberg. This is unverifiable, but what is known is that in 1876, Ben Wenberg, who traded in Caribbean fruit, demonstrated a new lobster-and-cream dish in a chafing dish to Charles Delmonico. His restaurant started making it under the name Lobster à la Wenberg. But soon angry words were spoken, the friendship was dissolved, and in a fit of vengeful dyslexia, Delmonico reversed the first syllable of "wenberg" to "Newberg."

In his 1894 book, *The Epicurean*, Charles Ranhofer, the chef at Delmonico's from 1862 to 1896, gave the following recipe for Lobster à la Newberg:

Cook six lobsters each weighing about two pounds in boiling salted water for twenty-five minutes. Twelve pounds of live lobster when cooked yields from two to two and a half pounds of meat with three to four ounces of coral. When cold, detach the bodies from the tails and cut the latter into slices, put them into a sautoir, each piece lying flat, and add hot clarified butter; season with salt and fry lightly on both sides without coloring; moisten to their height with good raw cream; reduce quickly to half; and then add two or three spoonfuls of Madeira wine; boil the liquid once more only, then remove and thicken with a thickening of egg yolks and raw cream. Cook without boiling, incorporating a little cayenne and butter; then arrange the pieces in a vegetable dish and pour the sauce over.

Milk thickened with flour is a technique that constantly reappeared in nineteenth- and early-twentieth-century cooking. Milk toast, a dish so bland that a personality type was named after it, was popular. This recipe comes from an 1853 book, *Miss Leslie's Directions for Cookery*, which is sometimes said to be the bestselling cookbook printed in America in the nineteenth century. Miss Leslie, born in 1787, lived most of her life in Philadelphia, except for a childhood in London, where her father managed an export company. Hers is another recipe that takes the precaution of boiling the milk first:

MILK TOAST

Boil a pint of rich milk and then take it off and stir into it a quarter of a pound of fresh butter, mixed with a small tablespoonful of flour. Then let it again come to a boil. Have ready two deep plates with a half a dozen slices of toast in each. Pour the milk over them hot, and keep them covered till they go to table.

This dish is sometimes thought to be a precursor of French toast, which the French call *pain perdu*, lost bread, because it is a way of using up stale bread.

In time gourmets became opposed to flour in sauces, which can cause a gloppy or pasty texture and add only a blandness to the taste. The fact is that a cream sauce can thicken through reduction without flour. The idea was not new, either. Charles Carter, a leading eighteenth-century English chef known to have some French influence but whose cooking often showed originality, wrote this recipe in his 1730 book, *The Compleat Practical Cook*:

Take good barrel-cod [salt cod], and boil it; then take it all into flakes, and put it in a sauce-pan with cream, and season it with a little pepper; put in a handful of parsley scalded, and minced, and stove it gently till tender, and then shake it together with some thick butter and the yolks of two or three eggs, dish it, and garnish with poached egg and lemon slices.

THERE WERE ALSO many eighteenth- and nineteenth-century butter sauces. Escoffier listed more than a dozen such sauces, some egg thickened like hollandaise or béarnaise, but many with butter bases such as lobster butter, shrimp butter, and *beurre noisette*, whose name does not refer to hazelnut butter but to color—it is a slightly browned butter. When the heat under *beurre noisette* is turned up, it becomes *beurre noir*, or black butter. *Beurre noir* can be heated slowly until it turns dark, and then an acid such as lemon juice or vinegar is added to stop the process. Capers are often added. In France the two most traditional plates served with this sauce are brains, either calf or lamb, and poached skate wing.

Here is Carême's recipe for the classic *raie au beurre noir*, skate with *beurre noir* sauce.

Take a skate, wash and brush it and cut off the wings, place [the wings] in cold salted water, bring to a boil and remove from the heat and let soak for a half hour.

Place a skillet over high heat, and put a piece of butter in it. When it begins to burn, toss in chopped parsley and two spoonfuls of vinegar and then cover the skillet to prevent the vinegar from evaporating.

Drain the skate, pull off the skin [if you grab it from the cut side it can be removed with a strong yank], put on the plate, sprinkle with salt, and pour the sauce over it.

Juliet Corson, a Boston-born social reformer and cooking instructor, tried to develop a popular cuisine for poor people through her New York School of Cookery and her books of inexpensive recipes. In her slim 1882 book, *Meals for the Millions: The People's*

Cookbook, she offers this less expensive *beurre noir* dish, "Eggs with Burnt Butter":

> Break half a dozen eggs, putting each in a cup to keep them entire; put four tablespoons of butter into a frying-pan and brown it over the fire, slip the eggs into the hot butter and cook them to the desired degree; then take them up with a skimmer, lay them on toast and set the dish containing them where they will keep hot. Pour half a cup of vinegar into the butter. Let it boil up once, pour it over the eggs, and serve them hot.

Another standard is white butter. *Beurre blanc* is made by cooking white wine, vinegar, and shallots together. As the story goes, it was invented by chance. Clémence Lefeuvre, a young nineteenth-century cook in the Loire Valley, where the kind of river fish that go well with a butter sauce are found, was preparing pike with a béarnaise sauce. This sauce, despite being named after the region of Béarn in southwest France, was a Parisian invention. Clémence supposedly started by reducing white wine, vinegar, and shallots, and then reached for egg yolks, which she meant to slowly beat into melted butter. But—oh, no!—she was out of eggs. She did have plenty of cold butter, though, and beat it into the white wine reduction, thereby making *beurre blanc* sauce. It is a good story for those who like to believe that ideas come about by accident. In any event, *beurre blanc* is traditionally served with white fish such as pike from the Loire.

MILK OR CREAM also started to replace flour and water in fish or clam chowder. At the end of the 1873 recipe for the fish chowder made by the Parker House in Boston were the words "If you do not find the chowder thin enough to serve well in tureen, add some fresh milk just before taking up and let it come to a boil." And by the end of the nineteenth century, New England chowder meant a milk or cream soup.

Cream soups became extremely popular in the nineteenth and twentieth century until, late in the century, cream and rich food began to lose popularity. In *Le Guide Culinaire*, published in 1903, Escoffier gave twenty-eight recipes for cream soups.

Rufus Estes, born a slave in Tennessee in 1858, became a highly regarded chef and, in 1911, the first black chef to publish a cookbook. Tomato soup made with milk or cream was an old standard, but Rufus Estes, true to his black Southern background, added green tomatoes:

Chop fine five green tomatoes and boil twenty minutes in water to cover. Then add one quart hot milk, to which a spoonful of soda has been added. Let come to a boil, take from the fire and add a quarter cupful butter rubbe[d] into four crackers rolled fine, with salt and pepper to taste.

The most famous cream soup was a New York invention, developed by the Frenchman Louis Diat, who moved to New York in 1910, became a U.S. citizen, and worked for forty-one obsessive, workaholic years as the chef of the Ritz-Carlton Hotel. Each summer he would invent a cold soup for the hot months in New York, and in 1917 he came up with a *crème vichyssoise glacée*, or iced vichyssoise cream. It is sometimes claimed that the soup was a French idea that he simply adapted, and Diat did say that the idea came to him from his childhood memories of the area near Vichy. And there, the French did make a creamed potato-leek soup, though it was usually not chilled.

I have to confess that this soup was one of my favorite treats as a child. I loved its presentation and taste, loved the way it was served in a metal bowl sitting on a dish of shaved ice, loved the way the bowl and the soup were so cold, loved the thick creaminess of the soup as I moved my spoon through it, and loved those bright green random dashes of chives. A Proustian thing, I can recall the taste as I describe it. There are many versions of this recipe claiming to be "the original," but this is how Diat gave it in his 1941 book *Cooking à la Ritz*:

    4 leeks, white part
    1 medium onion
    2 ounces sweet butter
    5 medium potatoes
    1 quart water or chicken broth
    1 tablespoon salt
    2 cups milk
    2 cups medium cream
    1 cup heavy cream

Finely slice the white part of the leeks and the onion, and brown very lightly in the sweet butter, then add the potatoes, also sliced finely. Add water or broth or salt. Boil from 35 to 40 minutes. Crush and rub through a fine strainer. Return to fire and add two cups of milk and 2 cups of medium cream. Season to taste and bring to a boil. Cool and then rub through a very fine strainer. When soup is cold add the heavy cream. Chill thoroughly before serving. Finely chopped chives may be added when serving.

MANY OTHER MILK and cream dishes also became popular. Here is Rufus Estes's recipe for a now forgotten milk dish, "baked milk."

Put fresh milk into a stone jar, cover with white paper and bake in a moderate oven until the milk is thick as cream. This may be taken by the most delicate stomach.

The French, probably the first to put milk in coffee, started doing so soon after the Turks brought coffee to Paris in the seventeenth century. But the Italian love of foamed milk and coffee began only in the twentieth century. In Isabella Beeton's influential 1861 *Book of*

Household Management, she did not write about milk served with tea or coffee for breakfast, but after her death, revised editions of her book kept being published, and in the 1890 edition were recommended menus for a week's worth of family breakfasts. Every one included tea, coffee, or chocolate with hot or cold milk.

Adding milk to chocolate had begun when the Spanish brought cacao back to Europe from the land of the Aztecs in the early sixteenth century and started serving hot chocolate drinks made with sugar and often other spices. Maria Theresa of Spain loved hot chocolate, perhaps her only solace while her husband King Louis XIV of France was sleeping with his many lovers, and it is said that the hot chocolate, or the sugar in it, was the cause of most of her teeth falling out.

Despite this seemingly negative celebrity endorsement, hot chocolate grew in popularity. Thomas Jefferson thought that it was only a matter of time until it replaced tea and coffee as America's favorite hot drink. In 1917 Alice B. Toklas and her partner, the writer Gertrude Stein, were in the south of France trying to help the wounded in Nîmes, and Toklas was so impressed with the hot chocolate there, which the "the Red Cross nuns in the best French manner served in large bowls to the wounded piping," that she wrote down their recipe:

> 3 ounces melted chocolate to 1 quart hot milk. Bring to a boil and simmer for ½ hour. Then beat for 5 minutes. The nuns made huge quantities in copper cauldrons, so that the whisk they used was huge and heavy. We all took turns in beating.

This was the real hot chocolate and milk, not the powder with all the cocoa butter removed, a process invented by the Dutch chocolate maker Casparus Van Houten in 1828. There are many commercial uses for the extracted butter. Today, it is hard to find a real hot chocolate. In 1974 the great American food writer James Beard wrote, "Chocolate now has become something that is tipped out of a little paper bag, into a cup, dissolved with hot water and served

with artificial whipped cream or a marshmallow stuck on top. This is
not hot chocolate and it really pains me to think that a whole generation is growing up never knowing the glories of a truly well made
cup of hot chocolate."

GIVEN THE WAY milk and cream were being used everywhere in the
nineteenth century, it was only a matter of time before bartenders
started adding them to cocktails. The word "cocktail," written then
as "cock-tail," was first seen in the *Morning Post and Gazetteer* in
London on March 20, 1798, leading some to believe it was of British
origin. The British have certainly always been cocktail enthusiasts.

In the Americas, cocktails became popular with the advent of ice
shipments to hot places like New Orleans and Havana, two cities that
became famous for them. Only later did drinks made with sweet
liqueurs and cream also became known as cocktails.

One of the first of these cream cocktails was the Alexander, made
with gin, chocolate liqueur, and cream in the late nineteenth century.
Then Rector's, a famous New York restaurant, started making that
drink with brandy instead of gin and grating nutmeg on top—the
brandy Alexander.

The grasshopper—equal parts crème de menthe, crème de cacao,
and cream—was invented in New Orleans. The two liqueurs were
already old drinks by that time, the mint from the nineteenth
century and the chocolate from many centuries before that. The
idea of combining the two with cream was credited to Philibert
Guichet, the owner of Tujague's restaurant, a New Orleans land-
mark since the nineteenth century. The earliest known mention of
it was in 1919.

Other dairy cocktails followed. The White Russian—coffee
liqueur, vodka, and cream or sometimes milk—was invented in the
Hotel Metropole in Brussels in 1949.

These dairy drinks had a reputation of being "ladies' drinks"
because they were sweet and low-alcohol. Aside from the sexist

assumption that women didn't like alcohol, not all milk drinks were low in alcohol. The Italian novelist Clara Sereni in her 1987 autobiographical novel with recipes, *Casalinghitudine*, loosely translated in the English-language edition as *Keeping House*, recalls one such "milk elixir":

1 quart milk
1 quart alcohol for making liqueurs [usually a high-proof white fruit alcohol]
2 pounds of sugar
3 teaspoons vanilla extract
1 lemon

I cut the lemon into small pieces (all of it, rind as well as pulp) and I put it in a large bottle together with all the other ingredients. It is better to dissolve the sugar first in some warm milk. I seal the bottle hermetically and leave it there for two weeks, during which time I simply shake it two or three times each day.

When the two weeks are up, I take a pan or bowl on which I put a colander in which I have set a tight-weave cloth napkin. I pour the contents of the bottle into the napkin-covered colander and leave it, as the filtering of the liquid is a lengthy process. The liquid will filter into the pan, and this of course will be poured into bottles and aged. A thick, yogurt-like cream, sweet-smelling and high in alcohol content, will remain in the napkin. This can be stored in small jars, and eaten with a spoon.

MOST OF THESE drinks, like cream sauces and soups, have faded in popularity as people have become more health-conscious and moved away from consuming foods with a high fat content. The exception was the White Russian, which experienced a reprieve from obscurity after being featured in the 1998 film *The Big Lebowski*.

The problem of harmful bacteria in milk has been conquered, but now it appears that dairy also contains other things that could kill you—such as cholesterol and fat. Eating heavy food has also gone out of fashion. And who would really want to have six slices of milk toast in a heavy milk and butter sauce for breakfast anyway? If you drank white Russians regularly, like the Dude in *The Big Lebowski*, wouldn't you become enormously fat?

— PART THREE —

Cows and Truth

Their attention is complete as they look across the road: They are still and face us.
Just because they are so still, their attitude seems philosophical.

—Lydia Davis, "The Cows"

15

THE BUTTERING OF TIBET

WHILE CHINA, WHICH for millennia has had no dairies, is trying to become the most modern and productive dairy nation in the world, northern Tibet—a country with a culture and Sanskrit-related language very different from China's—has had a strong dairy culture for centuries. In 1950, China took Tibet by military force and has been trying to assimilate it into Chinese life ever since. The Chinese have had some success in imposing both their Chinese architecture and the presence of Chinese people in Tibet's cities and larger towns. But out in the country, there is a different story. Nothing shows the difference between the Tibetan and the Chinese better than dairying. Observing Tibetan dairy farming offers a glimpse into medieval times

Xining is a fast-growing Tibetan city crammed with slender new skyscrapers, a strongly Muslim town with a lavish mosque. It is located in high mountain country, where some of the valley floors are almost two miles deep and lush with vegetation in summertime. Red-rock canyons have spectacular wind-carved pinnacles—like a land of banded red chimneys. In late spring, the banks of the nearby Yellow River are green, and the river is wide, rushing, fast-churning and, yes, really yellow. Climbing higher, the steep slopes are rich with green pastures in the spring and summer, brown and barren in the fall and winter. But even in the green time of year, the landscape is broken up in places by enormous sand drifts—it is as if a desert has been placed here and there. And in a sense, that is what it is: The sand drifts are part of the Gobi Desert of northern China and Mongolia. The Gobi was created when the rains were being blocked by the mountains of Tibet and the Himalayas beyond, and it is of some concern that the Gobi is now expanding into both China and Tibet.

High up in this country, where the air is too thin for trees and at times feels too thin for humans, is a land of summer grassland inhabited by mountain sheep, yaks, and nomadic herders. At the tops of high mountain passes, dozens of wild streamers in bright colors stretch across the roads or trails, or even from one peak to another. They are placed here by Buddhists to bring good fortune to travelers as they go through the pass. Small pieces of paper with an image of a horse, one of the oldest uses of wood-block printing, are thrown into the air by the nomadic herdsmen so that their horses will fly through the pass, carrying good luck with them. Sometimes, too, the herdsmen tie arrows together vertically with yak wool, and there are ritual fires burning yak butter, juniper branches, and roasted barley, one of the few crops that grows here.

The sheep of this country are not milked, because they yield too little. They are raised for meat—mostly mutton, though some rich people slaughter them when they are still small for tender lamb. But the main livestock and enterprise of these nomads is raising and milking yaks. The yaks are also combed for their wool.

Yak

Yaks hold their wide-horned heads below their great humped shoulders, making them look like huge, brooding, woolly, prehistoric creatures. They are bigger than cattle and tend to make guttural grunting noises. In Tibetan they are called *bos grunniens*, which means grunting ox. A typical bull is five feet tall at the shoulders and often taller. Wild bulls can be nearly seven feet at the shoulders. The Tibetan word *eYag*, from which "yak" comes, is used exclusively for domesticated yaks. Tibetans work with both domestic and the larger wild animals, sometimes breeding a domestic yak with a wild bull. A few other places in Central Asia and Mongolia have domestic yaks, but Tibet is the only place in the world with wild yaks, and the Tibetans' ability to breed the domestic with the wild variety maintains the domestic species.

Wild yaks are always black, and domestic yaks are usually black, too. But occasionally a white one is born, and these are particularly prized for their wool. The white ones, although bred with the herd, are also a different and larger subspecies than the black domestic yak.

Despite their bulk, yaks can move quickly, even at 20,000 feet. On rocky terrain they are as agile as goats, and even deep snow presents no obstacle to them. When a pass is closed by a heavy snowfall, the nomads need only to send a herd of yaks through to flatten the snow and open the pass.

The nomads wear clothes made of brightly dyed yak or sheep wool, and each subregion has its own variety of unique hats. Some of the hats look like cowboy hats, and some are much more elaborate, with fur and embroidery. But the use of fur on hats is fading—the nomads are Buddhist, and the monks have told them that it is wrong to kill animals.

The nomads are completely dependent on the yak. They use the dark yak wool not only for their clothing, but also for their homes, which are large pitched tents. The yak is also their main means of transportation and, with the exception of mutton, barley, and a few roots such as turnip that they grow in their lower fields, their food supply. Despite the admonition of the monks, they eat the meat, which they grill on a fire, and prefer it dried because it is easy to travel with when they move to better pastures. However, their primary foods are yak milk and yak dairy products.

The female yak, known as a *dri*, is not nearly as productive as a cow, but cattle can't survive in these high altitudes. Yaks thrive and reproduce at altitudes between 10,000 and 17,000 feet. Below that, they become sluggish and don't reproduce. Cows are usually inseminated at around thirteen months and give birth to their first calf when they are about two years old. Often yaks do not have their first calf until they are four years old, or in some areas, five, and unlike a cow, there are years when they have no calf. Nor do they produce a tremendous amount of milk for their size. They barely produce at all in winter, and it takes one thousand *dri* to produce about 130 gallons of milk.

Their milk is of extremely high quality, lactose-sweet, and 6 percent butterfat—far higher than most other milks, including the 4 percent of cows. With all that fat, Tibetans make a lot of butter. People who live in cold high altitudes need a high-fat diet. Yak butter is rich in omega-3 fatty acids and has a lush, pleasant flavor—or it would if it were refrigerated. The nomads don't have refrigerators, and the butter is allowed to go rancid, but they seem to be accustomed to the taste and smell.

Yak butter is sold everywhere and has many uses. It is stored in the dried ruminant belly from a sheep. This is thought to preserve it. Possibly it is preserved for a time, but it is usually used when it is about a year old, and by then it is rancid.

Entering a Tibetan Buddhist temple is, by design, mesmerizing. The temples are built high on hilltops, and unless you are used to high altitudes, you are light-headed and short of breath when you walk through the ornate front door. It is a world of saffron, gold, and red ornament, accompanied by the ringing of bells, the blowing of hoarse single notes on horns, and the rhythmic chanting of men in prayer. There is a smell—familiar, but stronger than usual. It is rancid butter. There are rows of bronze cups, each filled with yak butter with a juniper wick wrapped in cotton in the middle. Some of the cups are lit. All have a strong smell. The butter has been there for a year or more.

The altar is filled with elaborate sculptures that are three, four, or five feet tall. Their bases are usually coral-colored, with red ornamentation. Above, globes and various geometric shapes are carved in blue, green, and gold with black-and-white trim. Sometimes the sculptures display colorful floral patterns or mythical characters. All are made of solid yak butter, shaped with the same type of tools that sculptors use on clay.

The sculptures are created only by Buddhist nuns who are special butter artists living in an adjacent convent. They mix powdered dyes with the butter. It takes a nun one day to make a sculpture. They make them only three times a year, and they display each for a year before burning it.

Butter is a sacred food to the Buddhists because it is nourishment taken from animals without the use of violence. All the nuns and monks who live in the monasteries and convents are vegetarians, and they say that this is the Buddhist way. When it was recently pointed out to them that religious Buddhist herders in the high country eat mostly yak and sheep meat, a twenty-six-year-old nun answered, "We don't know anything about the outside."

THE NOMADS ALSO eat *tsamba*, which is the Tibetan national dish. It was created by nomads for nomads. To make it requires *chura*, the fresh curds produced in the first stage of cheesemaking. But since the Tibetans wanted curds that they could travel with, they dried them first in the sun, producing brown, crunchy little *chura* kernels. It has often been said that the swift-mounted Mongolian army ate cheese. But they did not eat a Western-style cheese; they ate dried curds similar to *chura*, packing them in their saddlebags for a quick energy snack.

Tsamba is made by pouring tea in a bowl and adding *chura*, barley flour, yak butter, and sugar. The ingredients are kneaded together by hand until the mass reaches the consistency of cookie dough, a bit shocking to Western sensibilities when it is made by a waiter at the table in a restaurant. To build very large temple butter carvings, they are stuffed with *tsamba* for solidity.

Tsamba is not only made by hand, it is eaten with hands. Hiking up the mountains and needing a protein boost, the nomad reaches into the bag on his yak and scoops out a blob of *tsamba*. Between the barley flour, the butter, and the sugar, it tastes almost like cookie dough, though it is not very sweet. Tibetans do not have a taste for sweets, or perhaps just haven't developed one because they don't have much sugar. Even their traditional sweets, called *juema*, are only moderately sweet. *Juema* are small brown cubes made from yak butter, *chura*, a grated local root, and only a very small amount of sugar.

It is also a Tibetan tradition to melt yak butter in a cup of tea, which would taste better if the butter weren't rancid. The more urbane Tibetans living in towns use fresh milk in their tea instead. It is believed that the Indian custom of drinking tea with milk, which was introduced by the Brits, originated in Tibet.

The best thing produced with yak milk is yak yogurt—the milk's high fat content makes for a very rich and flavorful food, with a thin

Tibetan yak cheese label. This yak cheese is made by Tibetan Buddhist monks, but it is not traditional and has not been a commercial success.

skin of leftover milk fat sitting on top. The nomads always eat their yogurt plain because they have no fruit to add. After eating the rich yak yogurt, cow yogurt can seem like a letdown—a fact business-people have noted. Yak yogurt is now becoming a successful commercial product in China.

The Chinese want to dismantle the Tibetan culture and make its people Chinese. In northern Tibet, this means getting the nomads to abandon their wandering life in the mountains and settle in towns. The Chinese have built numerous red-roofed, yellow-walled villages in which to settle the nomads, but they are completely empty. More successful has been the Chinese effort to stop the nomads from wandering by fencing off their ranging lands so that now they have only government-designated winter and summer camps instead of constant wandering.

But many of the nomads stubbornly remain on their high-altitude slopes. Many of their tents are now made of white canvas instead of black yak wool, though a few of the old wool tents are still in use. The new white tents often have solar panels that provide electricity for their inhabitants.

The nomads herd yak on horseback. Twice a day, the men gather the animals together on the steep slopes, while in the sun or pelting rain—often rain in the summer season when the grass is best—the women in their brightly colored dresses climb the grassy inclines carrying buckets. Laughing and in good humor, joking as they work, the women go from *dri* to *dri*, kneeling on their haunches, milking into buckets, sometimes wooden buckets, without using stools. The men stand idly by; milking is not their job. It is even said that a *dri* will not permit a man to milk her. But when a *dri* will not give milk, it is the man's job to bring over a calf to suckle for a minute to start the lactation. Then the poor unhappy youth is dragged away so the woman can milk the rest into her bucket. This practice was also common in Europe in earlier centuries.

The nomads immediately boil all their fresh milk. This has always been their practice, and despite the region's lack of refrigeration, there is no record of a great deal of illness or death resulting from consuming milk. There could be several reasons for that: Their children are generally breastfed, the cool high-altitude climate is less friendly to germs, and the milk is rarely consumed fresh. Most of the yak milk is turned into yogurt, and its fermentation would kill any dangerous bacteria.

Yogurt may be a greater force than the Chinese government in assimilating the nomads, because there is a growing and popular market for Tibetan yak yogurt. It is greatly appreciated by the Chinese and by Tibetans who have settled down and no longer make it themselves.

Some nomads have settled into a few hardscrabble towns that vaguely smell of mutton and rancid butter. Since they were built by the Chinese, they look Chinese, and like Beijing, they have extremely wide boulevards with overpasses for pedestrians. But unlike Beijing, there is almost no traffic and so people just stroll across the wide highways and rarely use the overpasses. The Chinese put up fences along the medians to discourage this practice, but the Tibetans find intersections or breaks in the fencing and cross anyway. It is the two cultures at an impasse.

The Chinese did make some attempt to imitate Tibetan architecture, and the walls of some blocks of houses have patterns painted on them. Occasionally, too, there is a Chinese ornament, such as an elaborate Chinese gate. But the principal attempt at aesthetic beauty is a habit of decorating buildings with garish neon, making the towns at night look like abandoned amusement parks.

In 2014, nomads Lha Zhongje and her husband settled in the town of Gabasongduozhen in the high mountain province of Tongde to open a yogurt shop. Here, they have a steady supply of yak milk, because their families are still nomadic herders in the high grasslands. They live in back of their small shop with their young son and daughter and sell products made by their families, including butter, dried yak meat, and fresh milk. But most of the fresh milk that they get they make into yogurt.

It is a simple process. On a hot plate in their home, Lha, dressed in a richly colored traditional dress, heats a large pot of milk. Then she adds a starter and stirs, covers the pot with a cloth, and lets it cool for three hours. Two kilos of milk makes 1.5 kilos of yogurt, which is their bestselling size. They have a refrigeration unit in which the yogurt will last for three or four days. Most Tibetans do not have refrigeration, and then the yogurt will keep for only two days. Lha and her husband sell about twenty tubs of yogurt a day and about 50 kilos, 110 pounds, of fresh milk.

"Yogurt is becoming more and more popular," she says. But there is a limit to their market, because most of their customers are ex-nomads who have moved to the towns, are struggling to find work, and have little spending money.

Sometimes the ex-nomads can find work on Chinese construction projects, but this is not regular employment. A lucky few find work with people like Droma Tserang and her husband, Ba Yo. The couple worked for twenty-two years in a Chinese state company that made yogurt, dried yak meat, and other regional products. Then in the 1990s, when the Chinese government started pressuring state corporations to be profitable, their dairy factory closed, as did many other state companies, and they and hundreds of thousands of others had no work. But the Tserangs knew how to make dairy products, and in 2006 they started making and selling yogurt and butter in their hometown. In 2013 they founded the Butterball Company, named after an odd habit that her mother had of keeping butter rolled in globes the size of melon balls. The Tserangs sold their trademark butterballs in see-through containers and also started making yogurt. Soon they got their first supermarket contract. By 2016 their company employed fifteen people.

AT THE AIRPORT in Xining, shops sell dairy products. The Chinese are convinced, not without reason, that these Tibetan products, even the ones from lower altitudes made with cow's milk, are far superior to anything people can buy in Beijing. The Tibetan milk, yogurt, and butter are sold in large cartons with convenient carrying handles, and on flights to Beijing, the overhead luggage racks are jammed with dairy products. This sight is a remarkable reversal of history.

16
CHINA'S GROWING TOLERANCE

CHINA HAS AN extraordinarily diverse cuisine and an ancient and revered gourmet tradition. However, throughout history, the Han, as the ethnic Chinese call themselves, as opposed to the Mongolians, Tibetans, and other ethnic groups living in the area, have rarely eaten dairy food. In fact, the consumption of dairy has been so rare that historically, many have assumed that the Chinese as a race were lactose-intolerant. What other explanation could there be for a milk reluctance not characteristic of other Asians?

Even the Japanese have a milk tradition. In the nineteenth century, a period of great modernization in Japan, the government greatly encouraged the drinking of milk, believing that it would create larger, stronger Western-type bodies. The military promoted this. The emperor Meiji insisted that he drank two glasses of milk daily. In 1876 the government established a large Holstein dairy on the northern island of Hokkaido, and even today, Hokkaido milk is famous in Japan as a high-quality product.

Mongols famously drank milk—mare's milk—and also traveled with their dried cheese curds and made *naisu*, little one-inch logs of dried milk that are chewy and slightly sweet. In 1123, in Feng Cheng, now known as Hu-He Hao-Te, the capital of Inner Mongolia, which, as opposed to independent Outer Mongolia, is now part of China, there was a street for cheesemaking named Lao Xiang, which means Cheese Road. It is still a street for making dairy products.

But China was different. The first mention of milk in China was during the Han dynasty (206 B.C.E. to 220 A.D.). This was the period when much of Chinese culture, including the Chinese language, was defined, which is why the Chinese call themselves the Han. The Chinese like to point out that the first mention of almost everything occurred during the Han period, because they were the first to document almost everything in writing. It was then that cow's milk was first recorded as being drunk in the imperial palace, and then that the first use of the Chinese word for milk, *niu lu*, appeared.

A few centuries later, in the fifth century, tea, which had been only a medicine, emerged as a popular beverage. The tea leaves were pressed into cakes, much like the *pu-erh tuocha* cakes made in Yunnan today, and boiled. Among the ingredients added to the tea were onions, ginger, salt, orange peel, pine nuts, and milk. But like most of the ancient Chinese milk traditions, this custom did not last.

According to Kwang-chih Chang, a Harvard anthropologist, there was a "continuous" introduction of dairy products into early Chinese culture, especially among the upper classes during the Tang period. Flourishing from 618 to 907, a dark time in Europe, the Tang Dynasty was a golden age in Chinese history, and among its gastronomic offerings was a milk-based frozen dish. Though not exactly ice cream, it was made from fermented buffalo's milk and seasoned with camphor, with flour added to give it solidity. It would have had a rather astringent flavor, and astringent flavors are still popular in Asian sweets today.

Writings from the Tang period praise the health-giving qualities of goat's milk. I Tsung, a ninth-century ruler, used to give his advisers "silver cake," a treat of which milk was a major ingredient. In South China, a starch called *sago*, made from the arenga palm and still popular in Southeast Asia, was mixed with buffalo's milk. In Szechuan, *shih mi*, "stone honey," was made from sugar and buffalo's milk.

Records show that Xuanzong, the first and, according to many, greatest Tang emperor, gave what was then considered an extravagant gift to General An Lushan. That gift was *ma lo*, also known as *koumiss*—the same *koumiss* that William of Rubruck and Marco Polo found the Mongols drinking centuries later. Like the Mongolian drink, Tang *koumiss* was also made from mare's milk, so it may have originated with the Mongols. They would skim the fat off the top of the fermented drink, and this they called *su*. A thick cream, it was used in extravagant dishes for the wealthy. But the greatest luxury of all was cooking with *t-i-hu*. This was made by heating the *su*, cooling it until it solidified, and then skimming the oil off the top. It was in effect butter, and very similar to the ghee of India. But while ghee was, and still is, a basic ingredient of Indian cooking, both *t-i-hu* and *su* were rarefied specialties, for only the very privileged. The great Chinese classical poet Pi Rixiu, writing of a lavish banquet, described the exquisite taste of a dish of swallow meat as being comparable to that of *su*.

From the Tang period on, the Chinese, like the Westerners, continually debated the relative merits of cow, goat, sheep, mare, and buffalo milk. In 1368, at the age of 100, Chia Ming was summoned before the new emperor. Asked what his secret of longevity was, he said it was eating and drinking carefully, and he gave to the emperor a copy of his book, *Yinshih hsu-chih*, "Essential Knowledge for Eating and Drinking." This is what he wrote about milk:

> Its flavor is sweet and acid. Its character is cold. Persons suffering from diarrhea must not consume it. Sheep milk taken along with preserved fish [the fermented sauce] will cause intestinal blockage. It does not go with vinegar. It should never be consumed along with perch.

This suggests that the Chinese struggled with milk upsetting their digestive system. And as Professor Chang has pointed out, none of the occasional uses of dairy in China ever caught on or became a feature of the cuisine. Was this because milk did not agree with people? Were they lactose-intolerant? Or was it simply because the privileged class kept dairy food to themselves?

THE REAL BEGINNING of the Chinese dairy industry occurred during the opium wars of the 1840s, which involved disputes about Chinese sovereignty and Britain's rights to trade in China. Jerseys and Ayrshires, then as now considered among the best British breeds, were brought into China to produce milk for the upper classes. They were called city cows because they were always farmed near cities, where the upper-class Chinese lived.

In the mid-1860s, though, the Chinese brought in cows to supply milk to the foreigners in Shanghai. Farmers who owned cows, either the Chinese yellow cow or the buffalo, brought them to town and milked them in the street to sell to foreigners. In 1870 the price of milk in Shanghai was fixed at one silver dollar for ten large cups. Only foreigners and a few wealthy Chinese could afford this price. In 1879 a Canadian missionary brought Canadian Holsteins to Nanjing, and the following year a British businessman brought Holsteins to Shanghai. The Chinese started to crossbreed their small herds with Holsteins. By the beginning of the twentieth century, small farms of five or six cows were established around the harbors and in all the major cities. In Shanghai, the first farm within the city was Yuan Sheng, which had ten cows. There was also a small farm just north of the Forbidden City, a neighborhood that is now part of

Chinese yellow cow

central Beijing. By 1945, the eve of the revolution, the biggest dairy in Beijing had forty-five cows.

In 1922, the College of Agriculture at the Canton Christian College in Canton offered a program in dairying and established an experimental dairy. The University of Beijing also created a teaching dairy. The Chinese were learning how to be dairy farmers. Textbooks were written, and one expert, Xu Fuqi, wrote prolifically on the subject.

IN THE CONTEMPORARY Chinese food world there is little awareness of China's hidden dairy history. Qu Hao, a popular personality on Chinese television food programs and the founder of a leading Beijing cooking school, told me, "There really are very few dairy dishes in China." He pointed out that even butter and yogurt are not used in Chinese cooking. But then he started thinking about exceptions.

"The best-known traditional milk dish is Cantonese, *daliang chao xiannai*," he said. This translates literally as "fresh milk from Daliang" but in English is usually called "fried milk." A starch made from beans, corn, or sweet potato is added to fresh milk, put in a pan with a little oil, and stirred over heat until it becomes a custard. It is traditionally eaten in a cup, though today shrimp or crab and a few vegetables are added to it. The original recipe probably goes back only to the late nineteenth century, so in terms of Chinese history, it is a modern dish. It is thought to have come from the Portuguese colony of Macao, where there is a similar dish.

Lin Hua, a Beijing food writer, told me about another Cantonese dish, *jiang zhuang nai*, which literally means "ginger bumping milk." She explained, "It is called this because the sweetness of milk acts on the pungency of ginger," and gave this recipe:

Crush ginger (or grate it), heat milk, pour it over the ginger, and add a little sugar. Heat into a custard.

She also pointed out that the Cantonese do have a few milk recipes, but only because of the long-standing foreign presence in the region. She said that the one milk tradition in Beijing is Old Beijing Imperial Cheese, which is actually not a cheese but a light milk custard. Her recipe:

Put milk in a pan. Add rice wine, the liquid from fermented sticky rice (sometimes some of the rice is also included, but usually just the liquid is used). Mix and bake for 40 minutes. Cool and refrigerate.

Hua then opined that Old Beijing Imperial Cheese is really just a Chinese version of Western custard. Furthermore, she added, "Today it is usually made in a microwave."

Old Beijing Imperial Cheese is sold in Beijing. Among its sellers is the Ma family, who own four small, clean, white shops, all just narrow rooms with a counter up front and a kitchen in the back. It is only a carry-out trade with no chairs or tables, but this is true of many Beijing prepared-food shops.

The Mas are Muslims, from the Hui people, a minority group that has a large population in Beijing. There are more than ten million Hui in China, and some Han are also Muslim. It has often been said that the Muslims are the only Chinese that eat dairy. The Ma stores are called Maji and have been in the family for generations. They live in Hebei Province, near Beijing, and once the location of the summer palace of the Qing emperors, where they ferment their own rice.

In the back of their shops they boil fresh milk with rice wine, cook it down until thick, add a little sugar, and pour it into plastic cups, which they then refrigerate and sell. The cheese is only slightly sweet and comes in plain, strawberry, blueberry, and taro.

The Mas also make other milk snacks, such as yogurt, brown crumbles of cooked-down milk scum, milk cooked to a soft dough and rolled with bean paste in the center, and milk cooked with almonds to make a light almond milk custard. They keep opening new stores because business is good and dairy food is newly popular in Beijing.

The Chinese are eating ever more dairy products, and fashionable young chefs are cooking a "fusion" cuisine with Western ideas incorporated into traditional recipes. This often involves using dairy products. Even Qu Hao, who is in his fifties and part of an older food establishment, confessed that he sometimes uses dairy products. When he makes the traditional white, fluffy, steamed buns known as *manto*, he adds a little milk to the batter. He likes the flavor it gives. And it makes the buns whiter. Today many Beijing cooks make *manto* this way.

CHINA HAS BECOME the world's third-largest milk producer, after the United States and India, and to Westerners it seems that this, along with many other Chinese economic gains, was an overnight success. But it took a long time to happen.

At the time of the 1949 revolution, China had a population of 500 million people and 120,000 cows, of which only 20,000 were Holsteins. The cows were producing small amounts of milk, and China was producing only 21,000 tons of milk annually. Powdered milk, mostly used for babies, was imported, with 90 percent coming from the United States, but it was extremely expensive, more costly than pork. Most Chinese women breastfed.

In 1953, in the totally state-run economy that was established postrevolution, 4,700 milk cows were under state management. Each cow produced only about 12 liters a day. But in 1957, the government decided to institute a dairy program as part of an overall plan to develop Chinese agriculture, and put its control in the hands of the army. Some said the program was a mistake because the Chinese people were lactose-intolerant. But the goal was never to provide everyone with milk. In 2017, the Chinese population numbered 1,386 million, so even if nine out of ten Chinese did not touch milk, that would still leave them with 139 million milk drinkers—more than the population of any European country. So the Chinese reasoned, not irrationally, that even if most people did not touch milk, they could provide it to a very large number.

By 1978 the government-owned dairies were working 480,000 cows and producing the equivalent of 1 liter of milk per year for every person in China. But fewer people than expected were buying it, because most did not have refrigerators. Refrigerators did not become common in China until the mid-1980s. By 2002, however, 87 percent of Chinese households had refrigerators, and between 1978 and 1992, milk production in China increased tenfold.

Close to ninety thousand cows were brought in to China from North America, Europe, and Japan between 1984 and 1990. China had planned to import many more, but the outbreak of mad cow disease stopped the export of cows from Europe and North America.

Now China imports cows only from Australia and New Zealand. New Zealand produces a Holstein that is slightly smaller than the usual one and eats less but produces well. In 2013, China imported eighty thousand cows and continues to import a similar number every year. A Chinese-Canadian project in British Columbia is providing China with Holstein embryos that are implanted in the native Chinese yellow cow. The Chinese have also started importing Swiss state-of-the-art milking equipment.

Today almost 40 percent of Chinese drink milk, the highest percentage in Chinese history. And while the Chinese are consuming more and more milk, the Americans are consuming less and less. Americans drink 37 percent less milk today than they did in 1970. Already, according to the United Nations Food and Agriculture Organization, the United States, long the world's leading milk producer, has fallen to second place behind India. It could in time fall to third. But China is not there yet.

Of the 40 percent of the Chinese people who consume milk, more drink imported powder than drink milk produced in China. In part, the rise in milk consumption is because breastfeeding has become less and less popular. Forty days after giving birth, most women have to go back to work, and breastfeeding becomes impractical. Also, many women want to appear fashionable and it is fashionable in China now to bottle-feed babies. Poor women who do not eat a rich-enough diet to produce sufficient milk also sometimes bottle-feed. This demand for milk for babies will probably increase since the government has dropped its one-child-per-family rule, and the birthrate is expected to rise.

The Chinese do not trust Chinese milk. That is why they carry it back from Tibet, or Australia, or New Zealand—wherever they go. In Beijing supermarkets, which are often enormous, they have aisles devoted to milk—mostly unrefrigerated UHT milk imported from New Zealand, though there is also refrigerated milk. UHT stands for "ultra-high temperature" and is a process of sterilizing milk by heating it to a very high temperature for only two seconds. A UHT carton that is not opened will remain unspoiled without refrigeration for up to nine months. UHT milk became popular in China in the days when there were few refrigerators. Some people don't like the taste of UHT milk, but many Chinese think it is safer. As Lao Li, the deputy general manager of a large private dairy, told me, "Food safety is the big issue."

Even in inexpensive markets for the poor, a considerable amount of space is devoted to milk. Organic milk costs twice as much as nonorganic milk, but people will buy it as well as the imported UHT milk, which costs even more.

Before 2008, people suspected that Chinese milk was full of additives that were not included in its labeling. And in 2008, everyone's worst fears were confirmed. Sixteen babies in Gansu Province were diagnosed with kidney stones. They had all been fed Chinese-produced powdered milk that contained a deadly industrial poison called melamine. Why would anyone add such a poison to milk? In quality tests, the melamine made the milk appear to have a higher protein content. In the following four months, three hundred thousand more babies became ill from the milk, and six of them died.

At first, the Sanlu Group, one of China's largest dairies, was found to be the source of the poisoned milk. But then other dairies in different parts of the country came under suspicion, and many believe that not all the culprits were found. Rumors of cover-ups abound. Many countries will no longer buy Chinese dairy products, and dairy imports are more popular than ever in China. In 2013, European supermarkets restricted sales of infant formula to two per customer because the supply had been depleted by Chinese tourists wanting to take Western formula home.

Leo Li manages the Wondermilk dairy, which was founded in 2006, before the scandal, by Charles Shao, a Chinese-American software engineer working for Google in Los Angeles. Shao had visited China and become convinced that there was no good, undiluted milk that people could buy with confidence. He brought Holsteins and Jerseys from New Zealand to China and established two dairies near Beijing and one in Shanghai. His original idea was to sell milk to the many foreign families who live in those two cities.

The dairy's name, Wondermilk, is designed to sound foreign. It is written in English on the front of the carton, and "won-der-milk" is written phonetically in Mandarin characters on the back. "People never believe that Wondermilk is Chinese," said Leo Li gleefully.

CHINA HAS A new and growing upper class concentrated in its cities, and, as has always been true in China, it is the urban rich who get the best milk. This is especially true because the wealthy crave everything Western—even the models in the advertisements of luxury products look Western—and dairy is Western. Ice cream is newly popular, and most of it is imported. Häagen Dazs, dubious umlaut and all, is popular, as are Western-style ice cream parlors.

Yogurt parlors are in, too. Yogurt is regarded as new and hip. One yogurt chain, headquartered in Xining, Tibet, and popular in Beijing, is called "Old Yogurt" in Chinese, but signs reading I LOVE

YOGURT in English are everywhere in the café. With the parlor's chic Western décor, the ubiquitous I LOVE YOGURT signs, and a menu offering yogurt parfaits with fresh fruit, yogurt cakes, milk candies, yogurt gelato, and smoothies, it is hard to remember that you are in Asia, let alone China. The only reminder is the parlor's excellent selection of Chinese teas. In 2016 there were seven "Old Yogurt" shops in fashionable areas of Beijing.

Coffee shops are also a new trend in China, started by Starbucks. The Chinese have not been coffee drinkers until recently. But now the affluent urban Chinese have a passion for cappuccinos and lattes. There is little interest in espresso—it is coffee with milk that everyone wants. Starbucks cannot open cafés fast enough, and there are many Chinese and Korean imitations.

Qiao Yanping, who has worked for the Chinese state dairy industry since the 1960s, said, "The Chinese now drink milk, but they don't eat milk," by which he meant that the one dairy product not popular in China today is cheese. This is the opposite of the situation in Japan, where cheese is very popular. There is one fast-growing cheesy product in China, however: The Chinese love pizza. Pizza Hut is opening store after store. Cheeseburgers are also becoming popular.

IF 40 PERCENT of today's Chinese drink milk, that is impressive, because it is estimated that only 40 percent of the world's people can digest milk. The other 60 percent are lactose-intolerant. So what happened to the supposed high level of lactose intolerance in China? One possibility is that it was always exaggerated. The fact that people do not consume dairy products does not necessarily mean that they are lactose-intolerant.

Li Cheng was a doctor in a Beijing hospital in the late 1980s when many patients started coming in with diarrhea and other typical symptoms of lactose intolerance. All these people had adopted the new fashion of drinking milk. Li Cheng and other doctors taught them to start by drinking only a little milk at a time and work their way up to more. Soon, they were drinking milk with impunity. Li Cheng and the other doctors contend that their patients' lactose intolerance had been caused by not drinking milk, and that by slowly reintroducing it into their diet, their ability to produce lactase had been restored.

Western gastroenterologists who have studied the condition find this highly unlikely. Having the production of lactase genetically shut down is the normal human condition and once it happens, they say, lactase production will never come back. An increasing number of scientists, however, believe that while this condition cannot change in an individual, it can, over generations, change in a population.

It has long been recognized that different groups of people have different dietary needs. It is also generally thought that one of the reasons for this is that humans evolved to suit their environments. Before animals were domesticated, most scientists believe, all humans had their lactase shut off between the ages of two and five. Since their mothers were their only source of milk, their need for milk had to be shut down so that mothers were not in a state of permanent lactation. Originally livestock was not raised to produce milk. But once humans started raising livestock, the genetic shutoff of lactase, through evolution, started to disappear so that humans could take advantage of the milk provided by their livestock.

There is a field called cultural genetics that studies how cultural changes cause the evolution of genes. In this field, it is thought that a population that needs and produces milk can over generations mutate the gene that shuts off lactase and the population can lose its lactose intolerance. This may be happening to the largest population in the world.

17

TROUBLE IN COW PARADISE

IN MANY HOMES in India, now the largest milk producer in the world, there are representations of two animals. One is of an elephant, the bearer of good fortune, whose image is always placed facing a doorway. The other is of a cow, the symbol of motherhood and therefore also of family and the joy of family life. This significance that Hindus see in cows has a simple logic: Cows give milk, milk sustains life, cows give life. But this is not to say that the use of cows and consumption of cow's milk in India began with Hinduism. In the south, in 2000 B.C.E., before Hinduism had reached that part of India, there were once large cow herds, apparently kept for both their milk and meat. It was very unusual at this point in history to have such a large quantity of milk production. As in other parts of India, their dung was also valued, as fuel. Archeologists have found huge mounds of cow dung ash in the south.

Indians get a great deal of milk from water buffalo, and in Kashmir people milk the *zomo*, a cross between a cow and a yak. Goats are also prevalent in the region. But in India, many believe that cows produce the best milk. In 1906, a popular nineteenth-century Bengali food writer named Bipradas Mukhopadhyay rated the milk of goats, sheep, camels, buffalos, cows, humans, mares, and elephants and concluded that while human milk was the best, cow's milk was second.

Mohandas Gandhi, the eccentric Hindu independence leader who ate almost nothing, did not agree. He drank goat's milk exclusively and was a great believer in its health benefits. He even took a goat with him to London in 1931 when negotiating with the British, which outraged his already infuriated racist adversary, Winston Churchill, but charmed his Indian followers.

For Hindus, the cow is sacred. The origin of this belief, like the origin of Hinduism itself, is uncertain, but the religion probably began in the Indus Valley in Northwest India. Around 2000 B.C.E., Aryan horsemen from central Asia invaded, bringing their religion with them. They worshiped a number of deities, and in the Indus Valley, that number grew. This, many claim, was the origin of Hinduism, which worships many deities, too, and is the oldest religion still practiced today. The early believers preserved their religion orally in hymns that were passed down through generations and became the *Rig-Veda*, which mentions cows seven hundred times.

There are some references in ancient Hindu literature to a ban on killing cows, but unlike some of their descendants, the Aryans did not have many bans or prohibitions, and seldom for religious motives. It is believed that the ban was originally economic, not religious. Cows were an important item of trade and functioned almost as a currency. Their religious significance seems to have developed later.

Lord Krishna, one of the best-known Hindu deities, is often depicted playing his flute among grazing cows and dancing milkmaids, known as *Gopis*. In fact, Krishna is said to have originally been a cowherd. He also goes by the names Govinda and Gopala, which literally mean "friend and protector of cows."

It was, and to a lesser degree still is, common practice for rural and even urban families to keep a cow for dairy purposes. Cows are still seen in the crowded, bustling traffic that clogs Indian cities, and it is considered a sign of piety for families to feed their cows before eating breakfast themselves.

Ghee, an oil produced from clarified cow butter, much like the *smen* used in North Africa, is a staple in Indian cooking, and is also burned during Hindu rituals. From the religious standpoint, ghee is sacred because it comes from a cow. Ghee is pure oil and so burns

This twelfth-century Khmer sandstone relief depicts the Hindu creation myth:
Vishnu creates the world by churning a sea of milk. (Musée des Arts
Asiatiques-Guimet, Paris © RMN-Grand Palais/Art Resource, NY)

very well, far better than the yak butter in the Buddhist temples of
Tibet.

Ghee can be heated to a much higher temperature than butter
without burning, and can be stored without refrigeration even in a
hot climate. Here is a very clear and simple ghee recipe from Mahdhur
Jaffrey, an Indian actress and popular food writer:

Melt one pound unsalted butter in a small heavy pan over a low
flame. Then let it simmer very gently for 10 to 30 minutes. The length
of time will depend on the amount of water in the butter. As soon as
the white milky residue turns to golden particles (you have to keep
watching), strain the ghee through several layers of cheese cloth.
Cool, then pour into a clean jar. Cover.

Indian farmers still make ghee, but most urban Indians buy ghee ready-made.

Both milk and butter are central to Hindu ritual. Traditionally, a few drops of ghee were sprinkled on rice as a purification ritual before beginning a meal. Kings were anointed with ghee, and princesses bathed in ghee. It was also used as a cosmetic to improve the complexion. Cow urine was once greatly valued and sipped during certain ceremonies, and cow dung, valued as fuel, was also smeared on the family hearth during certain purification ceremonies. The best purification material, *panchagavya*, was a blend of the five valuable products given by the cow—milk, urine, dung, curds, and ghee. Note that meat was not on this list.

Early on, restrictions were placed on drinking milk. Colostrum was forbidden. It was forbidden to consume the milk of a pregnant cow or a cow in heat or a cow that was suckling its own calf or the calf of another cow. This last restriction cost the dairy farmer a great deal of milk, and is anathema to most commercial dairy farmers.

The later religions of Buddhism and Jainism also have a special regard for milk, though the Jains, who must not kill, have to strain it through cheesecloth to make sure that there are no insects or living creatures of any kind in it. For Muslims, milk has significance as one of the special foods used to break fasts.

The gentle, reflective-looking, and you might even say soft-spoken cow was an excellent symbol for the Hindu ideal of a gentle, reflective-looking, and soft-spoken human. In ancient times, most Indians not only refrained from eating cattle, they seldom ate meat of any kind. Between the Hindu rejection of beef and the Muslim rejection of pork, India has the most evolved vegetarian cuisine in the world. But they are far from vegan. Their food, vegetarian as well as meat dishes, often involves dairy products. In fact, despite the dairy-obsessed Dutch, Swiss, and Scandinavians, there is no cuisine that uses more dairy than Indian.

Paneer is a simple, often homemade cheese from the Punjab region along the Pakistani border. A simple, fresh cheese, it is made

by adding acid to milk, which causes it to curdle, and then draining off the whey. Paneer is often cooked with vegetables such as spinach and often served with sauces that have other dairy ingredients. *Paneer Makhani*, cheese in a tomato cream sauce, is a Punjabi dish that has become an Indian classic. This recipe is from the *India Cookbook*, a one-thousand-recipe book by the food critic and historian Pushpesh Pant, who has tried to be the Ada Boni of Indian food and define this varied cuisine. An Indian cream sauce, even by nineteenth-century French standards, is very sophisticated. Note that here butter, not ghee, is used to create a rich sautéed cheese, and then vegetable oil rather than ghee is used for the frying to avoid using too much butter:

   3 medium tomatoes, chopped
   3 ½ tablespoons butter
   10 ounces paneer cut in cubes
   4 tablespoons vegetable oil
   1 tablespoon chili powder
   2–3 teaspoons ground green cardamom
   1 teaspoon Garum Masala [a very popular spice blend that can be
      bought preblended. It includes cumin, ginger, cardamom, cloves
      and several other spices]
   2 tablespoons ginger paste
   2 tablespoons garlic paste

1 tablespoon poppy seeds

2 tablespoons crushed ginger

1 bay leaf

3–4 cloves

2 cinnamon sticks about 1 inch long

1 teaspoon dried fenugreek leaves, crushed

1 teaspoon sugar

½ cup light cream

salt

4 tablespoons coriander (cilantro) leaves to garnish

Blanch the tomatoes by putting them in a large heatproof bowl of boiling water for 30 seconds, then plunge them in cold water. Remove the skin and chop the flesh.

Heat the butter in a heavy-based frying pan over medium heat, add the paneer and fry for about 8–10 minutes until evenly golden brown all over. Remove with a slotted spoon and soak in a bowl of water to keep succulent.

Heat the oil in a heavy-based saucepan, add the tomatoes and stir-fry for 2 minutes. Add all the ground spices, the ginger and garlic pastes and poppy seeds and continue frying 2–3 minutes or until the oil starts separating out. Add the bay leaf, cloves, and cinnamon and pour in 1 cup of cold water. Bring to a boil and cook over high heat for 5–7 minutes. Reduce the heat and simmer for another 3–4 minutes, or until the sauce is thick. Add the fenugreek, sugar, and cream, then stir, remove from the heat, and set aside.

Add the paneer to the sauce, season with salt and heat before serving. Garnish with coriander leaves.

Both vegetarian and meat dishes frequently had yogurt sauces. Cooking has always been the most popular use of yogurt in India. Before independence in 1947, India had about six hundred royal fiefdoms, each ruled by a maharaja or, in the case of a woman, a maharani. In the book *Dining with the Maharajas*, Neha Prasada compiled recipes from this aristocracy. This one comes from Kashmir in the far north. The dish is called *til ande ka achar*, eggs with yogurt sauce:

Eggs, hard boiled, halved lengthwise 5
Nepalese spice [Timur and Jimbu (these truly are from Nepal but
    are also popular in northern India. Timur is not a true pepper,
    but tastes like one. Jimbu is in the onion family and is a bit like a
    scallion.)] / a pinch
Cumin seeds ½ tsp
Whole red chilis 4
Sesame seeds 6 tbsp
Garlic pods [cloves] 4
Ginger [root] I inch
Salt to taste
Yogurt 14 ounces
Juice of lemon 2
Mustard oil I tbsp
Fenugreek seeds ½ tsp
Green chilies 4 [the long medium-heat Indian variety]
Turmeric powder I tsp
Red chili powder ¼ tsp
Green coriander leaves (cilantro) I tsp

Arrange the eggs on a dish.
    Heat the Nepalese spices and cumin seed on a griddle.
    Heat on a griddle, but separately, whole red chilies, sesame seeds, garlic, and ginger. Remove the skin of the garlic after heating.
    Grind the above spices in a mixie [food processor or blender; a coffee grinder would also work well]. Add salt to it.
    To the yogurt add the above spice mix, salt and lemon juice. [Add the lemon juice a small amount at a time.]

Heat the mustard oil, add the fenugreek seeds, green chilies, turmeric powder, red chili powder, and green coriander leaves: mix well. Remove and pour over the yogurt mixture.

Pour the above mixture over the eggs.

Though meat was often cooked in yogurt, it was sometimes cooked in fresh milk. Archana Pidathala, from the southern city of Chennai, collected the recipes of her grandmother, Nirmala Reddy, from the 1920s. This is her recipe for chicken drumsticks in a milk stew, *mumagakaya palu posina kura*. Chicken and lamb are the two most common meats in India, because both Hindus and Muslims eat them.

4 young tender drumsticks
½ tsp salt
I tbsp vegetable oil
½ tsp mustard seeds
4 garlic cloves peeled
10–15 fresh curry leaves [curry trees are native to southern India and
    the leaves can be bought at Indian spice shops]
I onion finely chopped
2 green chilies slit halfway though
a pinch of turmeric powder
salt to taste
I cup warm milk
2–3 tbsp chopped coriander leaves to garnish

Wash the drumsticks and lightly scrape the skin with a peeler or very sharp knife and cut into 2-inch pieces.

Boil 3 cups of water with ½ teaspoon of salt in a large vessel. And add the drumstick pieces. Cook on medium heat for 8–10 minutes or until the drumsticks are tender, but just short of fully cooked. To check whether it's just right, take a piece, cool slightly, press out a pod. Pop the pod into your mouth; it should be soft with a slight crunch. If cooked, drain the pieces in a colander and set aside.

Heat the oil in a deep, heavy-bottomed pan over high heat until very hot. Add the mustard seeds and when they sputter, add the garlic cloves and curry leaves, and sauté for a minute. Add the onions, green chilies, and turmeric powder and salt, fry for 3–4 minutes.

Once the drumsticks are cooked through, add the coriander powder and give it a stir. Now reduce the heat to low and slowly pour in the milk. Simmer on low heat for a minute, stir, and take the pan off the heat. Do not cook for longer at this stage, as the milk will curdle. Garnish with coriander leaves and serve with hot, steamed rice.

India also has many dairy desserts. This one, *phirni,* is from the Kashmir region and is included in Prasada's *Dining with the Maharajas:*

Milk 8 cups
Green cardamom 1 tsp
Almond slivers 24
Semolina, soaked in enough water to cover 12 tbsp
Sugar 9 tbsp
Cashew nuts ½ cup
Saffron mixed in milk ¼ tsp
Pistachios 2 tbsp

Boil the milk with green cardamoms. Add the almond slivers and soaked semolina; cook for twenty minutes, stirring continuously, so that the mixture does not stick to the bottom of the pan.

Add the sugar and mix well.

Grind the cashew nut with water in a mixie and add 6 tbsp of cashew nut paste to the above mixture.

Add the saffron mixture and pistachios. Transfer into a dish and refrigerate. Serve cold garnished with saffron, pistachios, and almonds.

Halwa is one of the most popular dairy desserts in India. It was originally Arab and was brought to India by the Moguls, a Muslim dynasty of Mongol origin who ruled much of India from their capital of Agra from the sixteenth to nineteenth centuries. Arab *halwa* had no dairy in it. It was the Indians who made it a dairy dish. The most popular *halwa*, an Indian standard, is *gajar halwa*, or carrot *halwa*, which was invented in the sixteenth or seventeenth century.

It was also in the sixteenth century that the Dutch, through the use of carotene, developed the first orange carrot. Previously, carrots were either pale yellow or purple, varieties that still exist. But the orange carrot was sweeter. According to popular food mythology, the orange carrot was developed in the mid-to-late 1600s to honor King William of Orange, but in actuality, it already existed before he was born. When the orange carrot appeared in Mogul India, there was great excitement. Vegetable *halwas* had started to become popular, and if there is one thing Indians truly love, it's bright, deep colors. Carrot *halwa* became, and has remained, a very popular dessert. This is Pushpesh Pant's classic recipe:

½ cup ghee
7 ½ ounces grated carrot
2 ounces heavy cream
½ cup sugar
2 tablespoons slivered almonds
½ teaspoon ground cardamom
I teaspoon rosewater (optional)
¼ cup chopped pistachios and slivered almonds to decorate

Heat the ghee in a large heavy-based pan, add the grated carrots and bring to the boil, stirring frequently. Reduce the heat and cook, uncovered, at a slow boil for 20 minutes, stirring frequently.

Add the cream and continue to cook, uncovered for I5 minutes, stirring frequently. Add the sugar and almonds and continue cooking for another 15 minutes, stirring frequently, until the mixture begins to stick to the base of the pan. Remove from the heat and allow to cool to room temperature. Stir in the cardamom and rosewater, if using. Serve decorated with chopped pistachios and slivered almonds.

The earliest known mention of *kulfi* is in the *Ain-i-Akbari*, a 1590 book on the administration of the sixteenth-century Mogul emperor Akbari. It gave a recipe for *kulfi* that called for a mass of *khoa*, a handmade concentrated milk. Condensed milk came later; today the milk is simply concentrated in the cooking process. Chopped pistachios and essence of *kesar* (saffron) are added. The mixture was frozen in a metal cone and sealed with wheat dough. The metal cone itself, the *kulfi*, was a word of Persian origin. Heavy cream, sugar, corn starch, cardamom, and sometimes dried fruit were also added. It is still prepared this way, still frozen in a metal cone, and sold on the street by *kulfiwalas*.

The British tended to sneer at *kulfi* and brought in machines to make European ice cream, but *kulfi* has always been popular with Indians. It is one of the oldest forms of ice cream. Though numerous writers point toward making it with condensed milk, most modern recipes specify fresh milk. The trick is to concentrate the milk. Here is Pant's modern *kulfi* recipe, a basic recipe to which various ingredients such as cardamom, mangoes, pistachios, or dried fruit can be added. The conical *kulfi* molds are easy to find in specialty shops or online:

3 ½ pints whole milk

¾ cup sugar

⅓ cup ground almonds

A few drops kewra water [a fragrant extract made from the male flower of the panderus plant, commonly used in cooking in northern India and in Arab countries and sometimes used in perfume]

Bring the milk to a boil in a large, heavy-based pan. Then reduce the heat and continue to cook, stirring frequently, for 45 minutes until the milk is reduced to less than half its volume. Remove from the heat, add the sugar and stir until the sugar is dissolved. Add the ground almonds and kewra water and allow to cool.

When the milk mixture is cool pour it into kulfi molds, leaving about 1 inch at the top for the frozen kulfi to expand. Seal firmly with lids and freeze for about 8–10 hours, or until frozen.

To serve, hold the cones under tepid water briefly, then remove the lid from the mold. Run a sharp knife around the inside edge and slip out the kulfi onto a plate. Kulfi must be served semi-frozen and not soft like ice cream.

Though *khoa*, which is boiled-down milk solids, is not used in *kulfi* anymore, it is still made and has many uses. Indian cooks give a great deal of attention to the different densities of boiled milk. (Nineteenth-century Americans, such as Rufus Estes, who had the recipe for a baked milk dish, also paid attention to boiled milk.) Milk boiled to half its volume is *panapaka;* to one third, *leyapaka;* to one sixth, *gutipaka*; and to one eighth, *sharkarpaka*, or *khoa*. South Americans boil down sweetened condensed milk to make *dulce de leche.* Condensed milk is a shortcut for making reduced milks because it is already *panapaka*.

Here is Pant's recipe for making *khoa* from 3½ pints of whole milk:

Put the milk in a kadhai [a pot for making *khoa*], wok, or deep, heavy-based pan and bring to the boil. Reduce the heat to low and cook, stirring after every five minutes, until the milk is reduced by half. Stirring constantly and continually scraping in the dried layer of milk that sticks to the sides of the pan, continue to cook until reduced to a mash potato consistency. Transfer to a bowl and allow to cool. When cool this paste can be stored in the refrigerator for up to 2 days. It can also be dried and stored as a solid: spoon the paste onto a clean piece of muslin (cheesecloth), then place in the

sink, weigh down with something heavy and leave to drain for about
1 hour. The resulting solid should be stored in the refrigerator and
can be grated or crumbled as required.

From Mahmudabad, once one of India's largest feudal estates, in
the north near Agra, comes a dish made with lamb and small balls
of *khoa*. *Khoa* is also used to make a number of milk candies. When
heavily sugared, it becomes *burfi*; and if cardamom is added, it becomes
pedas. Milk candies are extremely popular in India and Sri Lanka.

Milk candies are not unique to India. The British have toffee,
butterscotch, and cream caramels. In New Orleans, pralines are made
with milk, cream, sugar, and pecans. In the Philippines, there are milk
bars that are often made with condensed milk, and *pastillas de leche*,
which are made with buffalo milk.

This is a recipe for a milk sweet called *kalakand*, from the northern
state of Rajasthan, near Pakistan:

Ghee for greasing
2 ¼ pounds khoa, grated
1 cup sugar
1 tablespoon unsalted pistachio nuts, blanched and slivered

Grease a large baking sheet with ghee.

Heat the khoa in a kildhai [or a kadhai] or large, heavy-based pan,
then bring to a boil over medium heat, stirring constantly with a flat
spoon for five minutes. Remove from the heat, add the sugar and stir
until dissolved.

Transfer immediately to the greased tray and level with a spatula.
Sprinkle the pistachios over the top and thump the tray on a table or
on the floor to remove any air pockets. Leave it in a cool place to
set for 15 minutes, then cut into pieces.

Bengal, which straddles India and Bangladesh, is known for its
milk sweets. Traditionally, a Bengali meal always ends with candy

made from milk, and it is often served at weddings. Some Bengali candies are made from layered *sar*, the skins taken from the top of boiled milk.

Bengal is also known for its very old ritualistic drinks. Among them is *madhuparka*—a blend of ghee, curd, milk, honey, and sugar— which is offered to a student leaving to study with a guru, to a guest upon her or his arrival, to a seven-months-pregnant woman, to a bridegroom arriving at a wedding, and at the birth of a child or most any other important event.

TODAY, PASTEURIZED MILK is widely available and easy to find in any neighborhood "milk bar," which is a small dairy store. But throughout most of Indian history, a glass of safe, fresh milk to drink was a luxury item that only the wealthy could afford. And yet there have always been many milk-based drinks in India. Buttermilk was and still is extremely popular. I was invited to a cocktail party in Chennai, and at the last minute, the state government announced a dry day. Unable to serve cocktails at this elegant event at the city's best hotel, women draped in a dazzling array of shimmering saris served buttermilk in handsome earthenware glasses.

Lassi is another popular drink. Of Punjabi origin, the word "lassi" simply signifies yogurt mixed with water to make it drinkable. If a little sugar, pepper, and cumin are added, it is a *namkin lassi*, a salty lassi. Additional seasonings such as ginger, pistachios, blanched almonds, and green chilies make it a *lassi masalewal*, a spicy lassi. If just cardamom, rosewater, and saffron are added, it is a *meethi lassi*, or sweet lassi. Mangoes can be added to make a mango lassi, and strawberries to make a strawberry lassi. There are many lassis.

In the early twentieth century, the British promoted tea with milk and sugar as a way to get Indians to buy more black tea. To the Indians, tea with milk and sugar demanded the addition of spices. The most famous versions of these drinks are from the state of Gujurat in the far northwest, a state associated with Gandhi because it was his base during his years of political activism. *Elaichi ki chai* is tea with cardamom, sugar, and milk. But the most famous black tea drink, and one popular throughout India, is *masala chai*. "Chai," meaning tea, comes from a Chinese word by way of Persia. The word "tea" also comes from a Chinese word, but by way of the British.

Masala chai is usually made with black tea from Assam in northeast India, a region famous for its strong black tea. But in Kashmir, *masala chai* is made with green tea. The spices used to make the drink vary and can include black pepper, star anise, mace, chili, and nutmeg, as well as the ones mentioned in this classic Pushpesh Pant recipe from Gujarat. Note that some of the spice quantities here are given for the whole spice, which must be freshly ground. A coffee grinder does this well.

6–8 cloves, ground
8 green cardamom pods, ground
I cinnamon stick, I inch long, broken into pieces
I teaspoon ground ginger
I teaspoon ground fennel [seeds]
6 teaspoons [black] tea leaves
5 cups milk
¾ cup sugar, or to taste

Bring 6 ¼ cups water to the boil in a large, heavy-based pan. Add all the spices and boil for about two minutes over medium heat. Add the tea leaves and boil for further I minute, then reduce the heat and simmer for another 5 minutes. Pour in the milk and return to the boil, then reduce the heat and simmer for further 2 minutes. Remove from the heat and stir in the sugar. Strain through a fine sieve (strainer) into cups or mugs and serve hot.

Indians are tea drinkers. Only in the southern state of Tamil Nadu are the people predominantly coffee drinkers; drinking the local style of coffee and milk, which they call "filtered coffee." They use their own version of a double-decker coffee brewer, and after the coffee has dripped through the grinds to the lower section, they add boiling milk. By tradition, the milk is not homogenized, and as the boiling milk is poured into the coffee, the coffee maker skims off the top to use for making butter.

Milk punch is another milk drink that comes from the British. But punch itself is a a two-thousand-year-old Indian drink. The word "punch" is derived from the Sanskrit for "five," so called because it had five ingredients: alcohol (usually arrack, distilled from fermented palm sap), sugar, citrus juice, water, and spices. Some versions contained tea and no alcohol. Punch became extremely popular in seventeenth-century India among the many Europeans there—the British, Portuguese, and French—all of whom took their own variations back to their home country and other colonies. The British East India Company brought punch to England, and in the early eighteenth century, a punch made with milk, which resembled the possets that were popular at the time, came into fashion. It remained popular for only about fifty years, but a poem to "the West Indian posset" was written. Unverified rumor had it that it was penned by Alexander Pope, but if so, it was not among his best works:

> *From far Barbadoes on the Western main*
> *Fetch sugar, ounces four, fetch sack [sherry] from Spain,*
> *One pint, and from the East Indian Coast,*
> *Nutmeg; the glory of the Northern Toast,*
> *On flaming coals let them together heat*
> *Till the all-conquering sac dissolve the sweet.*
> *On such another fire put eggs, just ten*
> *(New-born from tread of cock, and rump of hen),*
> *Stir them with steady hand, and conscience pricking*
> *To see the end of ten fine chicken.*
> *From shining shelf, take down the brazen skillet,*
> *A quart of milk from gentle cow and fill it.*

When boiled and cold, put milk to sack and eggs,
Unite them firmly, like the triple league;
And on the fire let them together dwell.

DURING THE BRITISH Raj (1858–1947), as the time when the British ruled India is known, milk was supplied by a *gow-wallah*, a cow keeper, or sometimes a middleman. The cows were mostly local breeds called desi cows. *Desi* is a Sanskrit word that denotes native to the subcontinent. Desi cows are well adapted to the subtropical climate but do not produce as European breeds do. The yield from milking a desi cow might be a pint or even less.

In order to have an adequate supply of milk, the cow keepers tried to maintain huge herds of hundreds of cows. This is not unusual today, but in the days of hand milking, it was rare and extremely labor-intensive. The cow keepers also supplemented, often secretly, their cow's-milk supply with milk from whatever animals were available—goats, sheep, buffalo, camels, horses, and, according to some reports, pigs, though that was problematic in a country with a large Muslim population. The main substitute used was, and still is, buffalo's milk, even though Indians have always insisted that buffalo's milk impairs digestion.

But buffalo's milk has much to recommend it. Unlike cows, buffalos do not carry tuberculosis. Although it has reversed with modern methods, in the days of the Raj, a buffalo yielded more milk than a cow. Buffalo's milk also has the enviable quality of being lower in cholesterol even while higher in fat. It also keeps without spoiling for longer than cow's milk, a considerable advantage in a hot country without refrigeration. Also, buffalos are productive for up to twenty years, which is more than twice as long as cows.

Today, India is the largest producer of buffalo's milk in the world, producing more buffalo's milk than cow's milk even while professing to love cow's milk more. Numerous Asian and Middle Eastern countries, and a few African ones, produce buffalo milk,

Water buffalo

too. Small amounts are also produced in other places, such as the Campania region of Italy, where it is used to make mozzarella. Italy does not have indigenous buffalo—they were brought in Roman times, probably as draft animals. In the Philippines, buffalo's milk is used for making a cheese called *kesong puti*.

Buffalo's milk has long been a staple in India. In the fourteenth century, Ibn Battuta, the wandering Muslim, wrote while in India, "The buffalo here is in great abundance," and praised a porridge made with the milk.

Diluted milk from unscrupulous merchants was a common problem. To prevent it from happening, the wealthy had the *gow-wallah* bring the cows to their homes and milk them in front of them. Even so, there were stories of *gow-wallah*s having a little watered-down milk in the bucket before they started milking, and one story of a *gow-wallah* who kept a goatskin of water in his sleeve to squirt into the milk. Congee, or rice water, which is white, was often used to dilute milk as well. These practices led the British to keep their own private cow or cows at home, and many Indians did the same.

THE TOWN OF Anand became famous in the final years of the Raj because its local dairy farmers rose up against a huge private dairy company called Polson. Based in Bombay (now Mumbai), Polson had

established such a complete monopoly in Anand that the farmers
there had no other buyer to which to sell their dairy products.
Processing capacity had greatly increased, and the demand for milk
was up, but the British had arranged for all of this new profit to go
to Polson. The Patidar, a large peasant caste in the Anand region,
many of whom were militant nationalists, were angry and ready to
rebel, which made them easy to organize.

Sardar Vallabhbhai Patel, a Gujarat lawyer and one of the leaders
of the Indian National Congress that was fighting for Indian independence and later created the modern Indian state, saw the political potential of this fight. He organized a boycott, and the farmers
refused to supply Polson with milk. Then he arranged for a cooperative to collect the farmers' milk and ship it by rail to Bombay. In
taking these actions, he was following the lead of Mahatma Gandhi,
who had organized a 241-mile march across India in 1930 to protest
the British salt monopoly.

The management both of salt and of milk went to the heart of
the problem in pre-Independence India: the British were running
the Indian economy so that the British profited while the Indians
were plagued by famines. Millions were dying while the British
stood by. Since independence, there has not been a single famine in
India.

Called the Iron Man of India, Patel differed sharply with Gandhi
on certain issues. He believed that Gandhi's ideas about agriculture
were too romantic, and that for an independent India to succeed, it
had to develop technical and marketing skills for its agro industry.
One way to do this, he thought, was to establish cooperatives, and
in Anand, he saw an opportunity to demonstrate this idea.

In 1946, with no farmers providing milk to Polson, the British
government was forced to end its milk monopoly in Anand. It was
one of the defeats that made the British realize that they would have
to leave India. And Patel's dairy cooperative was given the hopelessly unmusical name KDCMPU, which stood for Kheda District
Cooperative Milk Producers' Union. It was India's first dairy

cooperative, and, understanding the importance of marketing, they gave it a more digestible brand name, Amul, for Anand Milk Union Limited.

Patel went on to be the first deputy prime minister of India and the minister of home affairs, but he died in 1950 and never saw how successful his cooperative movement became. After independence, the new India, charged with the task of economic development, made building dairy cooperatives one of its priorities. To that end, European breeds that did not do well in hot climates but had large milk yields were brought in and crossbred with the desi breeds.

Cultural attitudes also needed to change. In rural areas, working in the milk trade was often considered to be a very lowly activity. Milk was for the family, and someone who sold milk either lacked a family or was depriving them of their due. Milk products such as yogurt and ghee could be traded without such condemnation, however.

In 1949, Patel's dairy cooperative, now known simply as Amul, hired an American-trained engineer, Verghese Kurien, to manage its operation. Kurien was later dubbed the Father of the White Revolution—the movement that turned India into a major dairy power.

Using the model of Amul, the new Indian government established dairy cooperatives throughout the country as a means of efficiently producing and marketing milk. In 1970 the cooperatives received international support from the United Nations and the European Economic Community (forerunner of the European Union).

Since in India, as in many countries, cows were milked largely by women, it was initially assumed that many of the cooperative members would be women. But it soon became clear that while the cows were usually worked by women, they were almost always owned by men. The Indian government, realizing women's potential to expand the dairy industry still more, then developed women's dairy cooperatives. The first one was established in Anand, now known as India's dairy capital.

The government taught the cooperative women the newest techniques in breeding, feeding, milking, and general cattle care. They also selected some village women to be trained as managers and accountants. At first, men were used in these positions—and there were incidents of men stealing from the cooperatives—but as the women became trained, they took over and the cooperatives became completely women-run.

The women learned to reject the old belief that a cow should be starved during pregnancy because a well-fed cow could not carry a calf. In very poor communities, where people had trouble feeding their children, many hadn't understood the wisdom of feeding animals well. But the women in the cooperatives gained a different perspective and began growing not only food for their families, but also feed for their cows, including corn, sun hemp, and African guinea grasses, which prospered in parts of India. Some state governments, such as Tamil Nadu in the south, also helped the co-ops finance the purchase of European-Indian crossbreed cows.

The women did not love everything about the new cooperative life. Before, they had been ghee makers. A by-product of ghee making is buttermilk, and the women missed their once steady buttermilk supply.

The World Bank, which was helping to finance dairy expansion in India, favored European breeds. A crossbred cow, though short-lived, could yield four times as much milk per day as a native cow. The crossbreeds cost much more to maintain but were much less expensive to purchase. The cooperatives decided on crossbred cows.

Typically, a Holstein or a Jersey was bred with a Zebu. The large, humped, wrinkle-necked Zebu was an Asian breed that had been brought to India thousands of years before; it was the cow featured on the ancient seal of the Indus Valley. Zebus are unusually well suited for the tropics and have become popular in other hot countries such as Brazil.

By the twenty-first century, the cooperatives were so successful that private dairies started to reappear to challenge them. In 2017,

Zebu

the *Indian Express* predicted that soon private dairies would produce more milk than the cooperatives.

Meanwhile, people continue to keep a cow or two in the city, even though many metropolises ban this practice. Only cows that are associated with a religious site such as a temple are officially allowed. To feed their cows, people sometimes tether them in a park or public space. Nearby they place a tray of balls of fodder. The balls are for sale and a passerby can buy one, feed it to the cow, and then pull its tail for a blessing. In this way the owner is getting paid to have his cow fed.

The Indians also have many cow festivals. In Tamil Nadu, they celebrate Pongal, a winter harvest festival in which thanks is given to the sun. Cows are painted and their horns decorated. Indians love to add color.

And cows have remained a powerful symbol in India. In the late 1960s when Indira Gandhi formed a breakaway political party her symbol was a calf nursing a cow.

SOME ARGUE THAT the cow best suited for India is a Zebu crossbred with other high-yield local breeds. European stock is not suitable for the subcontinent, they say, as it requires too much medical care, the drugs used to cure it gets into its milk, and it still doesn't live very long. India is home to thirty-seven breeds of big-humped native cattle, all descendants of Zebus. But these breeds are dying out, while the population of the Jersey-Zebu crossbreds has increased 20 percent in this century.

Most observers also decry the many laws forbidding the slaughter of cows. Such interdictions have an ancient history. In the fourteenth century, Ibn Battuta wrote of a Muslim uprising against a Hindu ruler because he would not allow slaughtering. Then came centuries when there was no ban on slaughter. The issue began to reemerge in the eighteenth and nineteenth centuries, however, with the rise of Hindu nationalism. To favor a ban was to express anti-British sentiment. After independence, at the 1948 conference establishing the laws of the new Indian state, the issue of animal rights, and specifically cow protection, was raised. Was India to become the first country in the world to guarantee animal rights in its constitution? Many were enthusiastic advocates of the idea, but those focused on economic development opposed the move. Eventually a compromise was reached: The decision would be left up to the states.

The White Revolution progressed unimpeded by states. The economic development model of nationalism was more popular than Hindu nationalism. On November 7, 1966, there was a large march in New Delhi calling for a ban on cow slaughter. It was the first large demonstration against the Indian National Congress, the founding independence political party. The demonstration turned into a riot in which eight people died. Each year in Delhi, there is a march commemorating the event.

But the slaughtering of cows, strongly disapproved by some, continued. Since the 1980s, there has been a rise in support for the Bharatiya Janata Party (BJP), which endorses a kind of Hindu nationalism, a revival of Hindu culture and religion, and the rejection of what it terms European secularism. While not surprisingly very unpopular among Muslim Indians, it has strong support among Hindus. The BJP won control of a number of state governments, and in 2014 its candidate, Narendra Modi, was elected prime minister.

Those states with BJP-controlled governments instated bans on the slaughter of cows, and today, only eight of India's twenty-nine states have no restrictions on cow slaughter. Restrictions vary from state to state. Some states allow some slaughter but bar the selling of the meat outside the state. Export of beef from India is illegal. And the laws are not always obeyed. In April 2016, the *Times of India* reported that although slaughter was forbidden in the state of Uttar Pradesh, 126 slaughterhouses were operating there.

In a modern commercial dairy, which must carefully balance expenses against profits, a ban on slaughter poses serious problems. Cows, like humans and other mammals, do not remain fertile their entire lives and therefore cannot lactate for a lifetime. A cow, if well cared for, can live for many years after she stops lactating. For a large dairy, this could mean maintaining two herds, the first productive and the second completely unproductive. Maintaining a nonproductive herd represents a substantial financial loss. In other parts of the world, most dairy farmers ship cows that are no longer producing to a slaughterhouse. But in most Indian states, this is no longer allowed, creating a major problem, sometimes a crisis.

Some states, such as Rajasthan, have set up camps called *gaushalas* as homes for unwanted cows. Tens of thousands of cows are crowded into the *gaushalas*, where they are fed and cared for until they die. Such camps have existed since the seventeenth century, especially in Maharashtra, when they were used as a temporary relief system during periods of drought.

But there is more at issue here politically than simply animal rights versus the rights of farmers, a conflict that exists in many countries. The ruling BJP is seen by both Muslims and more progressive Hindus as anti-Muslim, and the laws restricting slaughter as anti-Muslim as well. Beef is the cheapest meat, and Muslims, India's poorest religious group, depend on it. Buddhists, Christians, and Sikhs also eat beef. In the fall of 2015 a mob dragged a fifty-year-old Muslim, Muhammad Akhlaq, from his home and beat him to death because he supposedly had beef at home, though his family said it was mutton. Such incidents of lynchings and beatings have continued. The press has labeled the perpetrators "cow vigilantes."

The ban is also regarded by some as a form of class warfare against the Dalits, the lowest caste in India, once called Untouchables. The Dalits have always done the things no one else wanted to do and used the things no one else wanted to use. They collected garbage, they scavenged for animal carcasses, they ate beef, and they slaughtered cattle for beef and leather. In fact, it was their association with dead cows that made them untouchable. The ban on the slaughter of cattle has decimated the economy and food supply of the Dalits, the poorest people.

Another effect of the ban is that it has turned farmers to buffalos. There are no restrictions on slaughtering buffalos. Buffalo's milk has a high fat content, and since milk is priced in India according to its fat content, it commands a high price. But it is not the most prized. The most prized milk is that of the purebred desi cow. A2 milk, the Indians call it.

India may be further proof that the more milk you produce, the more contentious issues will arise.

18

RAW CRAFTSMANSHIP

S INCE THE INDUSTRIAL Revolution, cheese factories have become bigger and more ubiquitous than ever. Some artisanal cheeses have disappeared. But many have endured and some of the ones that vanished earlier have been brought back.

Not only have artisanal cheeses survived, but the most famous ones of the preindustrial age remain among the most celebrated. While cheese has always been a common food, it has also always been revered by gourmets. Jean Anthelme Brillat-Savarin, an eighteenth-century politician, lawyer, and early food writer, famously declared, "A dessert without cheese is like a lovely lady with only one eye."

Brillat-Savarin insisted that the greatest cheese was Époisses, while Grimod de La Reynière, France's first great food writer, put Roquefort at the top of his list. He called it "the drunkard's biscuit" because it induced thirst. According to Grimod, the chief function of cheese was to lead to the enjoyment of wine.

As it happens, Époisses and Roquefort are the two favorite cheeses of this twenty-first-century writer. Neither King Louis XVI nor the statesman Charles Maurice de Talleyrand-Périgord, briefly prime minister of France from July–September 1815, agreed. They were both passionate about Brie, and it was said that the king was in the process of enjoying a fine Brie when he was arrested and then executed, a dubious celebrity endorsement. Talleyrand was a tireless promoter of Brie, which he called the king of cheeses, supposedly at a dinner at the Congress of Vienna in 1815, when Europe was slicing up post-Napoleonic Europe. It was said that Brie was the only king to whom he always remained loyal.

In 1896, Oscar Tschirky, the Waldorf-Astoria's chef, named Brie from the Paris region the most popular French cheese and Camembert from nearby Normandy the second most popular. Both of these cheeses have many imitations, both factory-made and artisanal, around the world. But the real Brie or Camembert are made in very specific places and shaped by these places, in the same way that only a wine from Burgundy is going to taste like a Burgundy.

A top-quality cheese is made shortly after milking, when the milk is considered to be at the optimum temperature for curdling. A true Brie is made from the raw milk of a specific area of Île-de-France. A Brie-de-Meaux comes from one place and a Brie-de-Melan from another. Raw milk has a distinct flavor, derived from the specific grasses in the pasture where the cow is grazed. The milk for a good Brie all comes from the same pasture. But the grasses change from one season to another. A fall Brie-de-Meaux does not taste like a spring one, and fall and spring Époisses are very different from each other. A good *fromagerie* in Paris lets its customers know the season of its cheeses. All of which is to say that a true artisanal cheese can only be made in small batches on a farm.

Artisanal Brie is made from the milk of a Belgian breed, the Belgian blue, a gray, lean, and extremely muscular cow. Its milk has

Belgian blue

less butterfat than does the milk of the white-and-black-speckled Normandy cow, whose milk is used to make Camembert.

Though Cuba is not well known for cheesemaking, the most famous Camembert imitator was Cuban leader Fidel Castro. Aside from his personal love of eating ice cream and gulping chocolate milk-shakes at the Habana Libre, he believed that building up the Cuban dairy industry was an essential priority.

Milk and milk products were an important part of the Cuban diet, and before the revolution, much of it came from the United States. A subtropical country, poorly suited for cows, Cuba had never produced milk efficiently, but after the U.S. embargo was imposed on the island country in 1962, Castro was determined to improve Cuban dairy farming. In time this would lead to the invention of Cuban Camembert.

In the 1960s, most Cuban cows were descendants of either the Spanish criollo cow or the Zebu from India. Both of these cows, though well adapted to hot climates, produced little milk. Castro's solution, like the subcontinent Indians', was to import Holsteins. They would be kept in air-conditioned stables. Given fuel costs, air-conditioning cow barns was not a commercially viable way to produce milk. But in those days, the Soviet Union was subsidizing the Cuban experiment, and dairy was a priority. The Cubans imported thousands of Holsteins from Canada.

A third of those Canadian cows died in a matter of weeks, leading Castro to pronounce that they would have to invent a new Cuban breed, "the Tropical Holstein." This new cow ended up being the exact same cow as the one developed in the Indian cooperatives—a cross of either a Holstein or a Swiss Brown with a Zebu. Most of these cows did not do well in Cuba either, probably because of a lack of proper diet. Holsteins, even half Holsteins, need to be well fed.

But there was one exception, a cow named Ubre Blanca, which means "white udder." This cow produced a phenomenal amount of milk, so much so that it made it into the 1982 *Guinness Book of Records* as being the cow that held the world's record for producing the most amount of milk in a day—109.5 liters. As well as the highest yield from a single lactation cycle—24,268.9 liters. The cow was a hero of the revolution, frequently reported on, and when she died in 1985, her official obituary said, "She gave her all for the people." The Cubans never produced another cow that made a comparable contribution.

For a while, the Cubans tried to produce cows small enough for people to keep in their homes for their own dairy needs. Nothing worked, but the Commandante was feeling good because he had come up with an idea for a star cheese product. He had decided that he would make the best Camembert in the world. By most accounts, Cuba was already producing a pretty good Camembert at that time, but he wanted something even better and instructed his cheesemakers to come up with it.

A short time later, in 1964, when the revolution was still young and of boundless ambition, Fidel invited André Voisin to Cuba to give a series of lectures. Voisin was a French war hero who had developed a theory of "rational grazing" that was influencing farming practices all over the world. Fidel asked him to sample the new Cuban Camembert, and the Frenchman said it was very good, like a French Camembert.

But Fidel wanted more and tried to get him to say it was *better* than a French Camembert. Finally, a frustrated Voisin grabbed the cigar from Castro's shirt pocket, a big Cohiba, the type of cigar developed by the Castro regime to replace the ones made pre-revolution. The Cohiba was generally considered to be the world's best cigar. Voisin asked Castro if he thought the French could make a better cigar than this.

Some days later Voisin died of a heart attack in his hotel room, and he is buried with many of Cuba's greats in Havana's Colón Cemetery. His memory is revered in Cuba despite his refusal to endorse their Camembert.

THE WORLD WANTS more cheese. That's why cheese factories were started. The world's population keeps growing and the amount of

Frenchman and women at work in a dairy, ca. 1910. (Adoc-photos/Art Resource, NY)

cheese eaten per person keeps growing. In an economic system that disfavors small-scale farming, where small farms fail every day, artisanal cheesemakers could never produce all the cheese people want. In France alone, the per capita consumption of cheese increased more than five times between the 4.4 pounds per person per year in 1815 and the 23.3 pounds in 1960.

Holland has almost completely ceded its famous cheese tradition to factories. Few people today have even tasted Dutch farm cheese, *boerenkaas*. In the most famous Dutch cheese district, Gouda, there are still a few stubborn farm families making farmhouse Gouda, but not many. The hard and nutty dark handmade Gouda has almost nothing in common with the factory-made Gouda known to most of the world, except for its shape.

Farmhouse Gouda, traditionally made by women, is produced with the raw milk of cows that graze in the same pasture. Their milk is brought in twice a day after milking, and the cheesemaking process is started while the milk is still warm. The first step is to sour the milk with lactic acid. Today this acid can be purchased, but only two generations ago it was still made by the women. Getting the lactic acid exactly right was one of the keys to making a good Gouda. When finished, some of the farmhouse cheese is sold in the local market, but most of it is sold to cheese merchants. It has been done that way since the fourteenth century.

Today, only an estimated 1.5 percent of the cheese made in Holland is artisanal, and there are only about five hundred such artisans left.

ALMOST ALL OF the most famous internationally known cheeses—Brie, Camembert, cheddar, Parmesan—are made from cow's milk. In this upper stratum of world-famous cheeses, there is one exception, Roquefort, which is made from sheep's milk.

The village where Roquefort is made, Roquefort-sur-Soulzon, has fewer than two hundred inhabitants. Only cheese that is made in this small village, which sits along a narrow road on a ridge beside a massive outcrop of rock, is allowed to be called Roquefort. This makes the real estate here too valuable to be used for anything else. The monopoly on the name was granted by King Charles VI in 1411. In 1961 a court ruled that even if the ingredients and technique used to make the cheese were followed exactly by cheesemakers elsewhere, only cheese that matures in the natural caves under the village, the caves of Mont Combalou, can be called Roquefort.

This is not without reason. Inside the Mont Combalou caves, the circulation of air and the humidity trapped in the rocks creates a unique mold-growing environment. The natural cellars are four stories below the village, a hundred feet or more deep into the rock. It is cold and humid in the caves, and the rock walls, the old hand-hewn wooden beams, and the wooden shelves where the cheeses are aged are all wet and slippery. The rocks show a kaleidoscope of colorful molds and lichens, and this is essential to making Roquefort Roquefort. The cellars are centuries old, and attempts to create new cellars always fail.

In the beginning of the 1990s, health inspectors from the European Union visited the caves and were appalled. Food was being left on moldy antique wooden shelves. Imagine the possibilities for dangerous bacteria. The cheesemakers argued that bacteria couldn't live in Roquefort because it contains live penicillium. Nevertheless, they were ordered to replace the wooden shelves with hygienic plastic ones.

Slowly and reluctantly, the cheese companies made the shift. All except Jacques Carles, who today is one of only five small independent Roquefort producers left. Most Roquefort today is produced by two giant companies, Société and Papillon.

Carles had inherited his company from his father, who started it in the 1920s when there were a dozen such small Roquefort companies. A traditionalist, he ignored the European Union edict and did not remove a single shelf. This saved him a great deal of money once it was discovered that Roquefort did not taste the same on plastic shelves and the old wooden ones were put back.

The village of Roquefort is in one of the wilder, least settled parts of France known as Saint-Affrique, in the Aveyron *département*, an area just south of the Massif Central. The people here speak their own dialect of French, and anyone who arrives speaking standard French is greeted with suspicion for fear he or she is a government inspector from Paris or, worse, a representative of the European Union in Brussels.

The villagers have their own ancient ways of doing things, and their response to anyone who wants them to change is that they should not argue with success. The one-street village is packed with trucks loading up to take their cheese around the world and the town smells as much of money as of blue cheese.

There are many rules regarding the use of the name Roquefort. Milk must be delivered within twenty days of lambing; the sheep, a relatively productive milker called Lacaune, must be from the Aveyron and a few neighboring areas; and the penicillium must be produced in the caves of Roquefort-sur-Soulzon. There are differences between the small cheese producers and the big companies.

The small producers collect their milk in old-fashioned cans, not tanks. The curds are stuffed into molds by hand. The bacteria, penicillium, causes the fresh cheese to bubble and in time the bubbles turn blue. It used to be distributed from powdered bread, but today a liquid version of the exact same mold from the caves is used. The producers swear it is the identical mold, and the EU accepts this, though it is a violation of one of the fundamental tenets of what makes a cheese a Roquefort.

Traditionalists still use bread to culture the penicillium. Carles buys a rye-and-wheat bread from a baker in the nearby village of Plaisance. He lets the bread mold in the caves, cuts off its crust, and then turns it into a blue powder.

The milk for making the Roquefort comes from more than two thousand farms. It takes a great many sheep, which are notoriously poor milk producers, to make cheese, which is why there are so few leading sheep cheeses in the world. The all-white Lacaune sheep has a reputation for producing a good amount, but a one day's milking of twenty sheep produces only enough to fill a 40-liter milk can. Some of the sheep are still milked by hand, but most are milked by machine. And the sheep farmers are paid well for their product. A farm with a good-sized flock can earn better than $100,000 a year from milk.

Like most artisanal cheeses, Roquefort is a highly seasonal product. From July to December, no milk is collected, because the

Lacaune sheep

ewes are allowed to feed their lambs. During this period, the cheese-makers sell what they call the *vieux fromage*, old cheese. In December they begin making cheese again, but this cheese, *nouveau*, will not be ready for sale until spring. From April until June both old and new cheese are available. And both have their fans; the *nouveau* is subtle but complex, and the *vieux* is strong and nutty. But none of this is marked on the label. As with Époisses and Brie, most customers don't know the differences unless they are local or buy from a very good *fromagerie* in Paris.

Roquefort cheese is made from unpasteurized milk, another one of the requirements to carry the Roquefort name. Many countries do not generally let raw milk products come through customs. But Roquefort is an exception. It would be unthinkable not to have Roquefort.

AT THE OTHER end of the Pyrenees is another much-loved but much less famous sheep cheese, made by the Basques, who have fought hard for every inch of their identity and are therefore careful about what they call Basque. Truly Basque cheeses, hard and pungent, must be made only from sheep's milk. The French government designates the Basque cheese Ossau-Iraty and allows it to be made with various types of milk. But the Basques call it *ardi-gasna*, sheep cheese, and stipulate that it must be made not only with sheep's milk, but with Basque sheep's milk.

The Basque have a Basque everything—the Basque sheep, the Basque pig, the Basque horse, the Basque dog, the Basque goat. The Basque sheep is the *burru beltza*, "black face," known in French as the *tête noire*, "black head." It is a white sheep with a black face and is unique to the Basque region.

Basque sheep are a bit difficult. They do not always behave like domestic animals and can seem wild. Sometimes they are difficult to milk. Also, they do not produce much milk. Crossbred Basque breeds, such as the *tête rousse*, "red head," and Basco-Béarnaise (bred with sheep from neighboring Béarn) are easier to handle. Neither of

Burru beltza *(black face), in Euskera, the*
Basque language. (Print from a woodcut
by Basque artist Stephane Pirel)

these produces great quantities of milk either, but their milk is
known for its flavor.

What is not acceptable, at least to the traditionalists, is using cow's
milk, goat's milk, or even the milk from the Lacaune sheep from the
Roquefort region when making Basque cheese. Nonetheless, some
Basques have been bringing in the Lacaune because a Lacaune will
produce roughly twice as much milk daily as a Basque sheep. Still,
the true Basque connoisseur wants only true Basque cheese made
from true Basque sheep. The only question is whether he or she
prefers the sharper Basque cheese from the scrubby high country or
the milder Basque cheese from the low pasturelands. This also is
never marked on a label, and the customer has to either know which
farms are where or talk to someone who does.

Jean-François Tambourin and his sons Michel and Guillaume
make real Basque cheese on their farm, Enautenea. The farm is at
the foot of Jarra, a velvet green slope in the rugged foothills leading
to the Pyrenees. The high pasturelands here have slopes pitched at
sixty degrees or steeper. The Basques joke that Basque sheep have

Basque Cheese label.

one pair of legs shorter than the other so they can stand sideways on the slopes. These green pastures with purple crests behind them, this land near the town of Saint-Étienne-de-Baïgorry, in a narrow mountain pass between France and Spain, must be among the most beautiful pastures that livestock anywhere have ever seen.

Only family members work the Enautenea farm and only Basque, Europe's oldest language, is spoken here. In school the children learn French and get French names. It was in school that Frantxua became Jean-François.

The family knows what they have. Jean-François, a lanky man with a thin weathered face and long Basque nose, waves his arm in a grand gesture of awe at the rugged crest, steep green pastures, thick dark woods, and terraced vineyards around him. "It's an inspiration to see the mountains, the woods, the vineyards," he says. The family has operated the Enautenea farm since at least 1788. They are one of ninety milk producers in the area.

They work with three breeds of sheep: the tête noire, tête rousse, and Basco-Béarnaise. They also raise Basque pigs, which, unlike the sheep, are a gentle, docile breed because their huge ears serve as natural blinders, cutting off peripheral vision.

The family maintains a herd of three hundred sheep. Each November they get four hundred lambs. They keep the seventy best females and the one best male. The rest are sold for meat. The sheep graze wild on the high mountain slopes, eating the grasses and high-protein clover that help them to produce milk. In the winter the family supplements their feed with grains grown on their farm or other local farms. They harvest the grains and grasses in summer to use for winter forage.

Farms with Basque sheep do not produce milk all year. They could if they fed the new lambs rather than letting them suckle, but the Basque believe that it is important "to respect the seasonal nature of sheep." So at most, the sheep produce milk 265 days a year, having summers and early fall off. Jean-François and his family, on the other hand, have no days off. Jean-François hikes an hour up the mountain every day because the flock has to be kept together. If not, the sheep become agitated. Otherwise, "they live in the wild, feed off the land, and usually don't need feed or medicine," he says. The sheep also have to be milked by machine twice a day. Before 1992, they were milked by hand.

The sheep's milk is never pasteurized. A group of experts tests it every month, not only to make sure it is untainted, but also to access its fat content, flavor, and color.

Most people in the world don't know what Basque cheese is or what the difference is between a factory-made Basque cheese, a cow's-milk Basque cheese, and an authentic sheep's-milk Basque cheese. But the Basque people know, and it matters to them.

GREECE IS KNOWN for goat's and sheep's milk, and there is a reason for that. The country is mostly arid and rocky, with very little grassland. The most important cheese in Greece is feta, which can now be made in a factory from any combination of goat, sheep, and cow's milk, and with either raw or pasteurized milk. But of course, feta was originally made from raw milk, from either goats or sheep, and those trying to make true feta today still use raw milk—usually 30 percent from goats and the rest from sheep. Seventy percent of the cheese currently consumed in Greece is feta of one kind or another.

Scientists have been studying Crete to try to understand the low incidence of heart and circulatory disease there despite the fact that 40 percent of its people's caloric intake is from fat. What makes this especially interesting is that almost all of that fat comes from olive oil and feta.

Feta is probably one of the oldest cheese types. Farmhouse feta was, and still is, made from the milk of free-range goats and sheep, curdled while still warm with the natural rennet from the stomachs of sucking animals. The curd is ready for cutting within an hour and is then placed in molds—traditionally, basket molds, but today metal ones, which are more hygienic.

Another hour later, the cheese is ready to drain and salt. It is then cured in brine made from sea salt and left to ripen. On the island of Limnos, people cure feta with salt by leaving it in seawater.

Greece has about four thousand dairy farms that provide milk for feta, and the typical herd numbers three hundred animals. The European Union says that in order to be called feta, the cheese has to come from one of six regions of Greece: Macedonia, Thrace, Thessaly, central mainland Greece, the Peloponnese, and Lesbos. There is a noticeable difference among the cheeses of the different regions. Feta made in Thessaly and central Greece is more flavorful. Peloponnesian feta also has a lot of flavor but is drier and more crumbly. Cheese from Macedonia and Thrace is milder, creamier, and less salty and has fewer holes.

There are about fifteen hundred cheese dairies in Greece that make feta in this traditional way, and they are all located near dairy farms so that they can start making the cheeses immediately after milking. A few have succumbed to pasteurization, and though they use low-temperature pasteurization to do as little damage to the milk as possible, aficionados say that this feta is not as flavorful.

The traditional feta makers are all competing with the feta factories, most of which are located in Europe and the United States. The factories use mostly pasteurized cow's milk and industrial techniques. They need only 5.3 liters of milk to make one kilo of feta, whereas the Greek farmers need 7.3 liters. The factory-made feta does not really resemble the Greek handmade, but most customers have never tasted the original.

Though feta dominates in Greece, there are also seventy other Greek cheeses, most of them artisanal. Almost every Greek island and peninsula has its own cheese, usually made from sheep's or goat's milk.

In the Cyclades, a group of Aegean islands that includes some of Greece's most popular tourist spots, are three less visited islands— Tinos, Syros, and Naxos. Venetians came to these islands in the fifteenth century and didn't leave until the seventeenth. They left behind their Roman Catholic religion (whereas most of the rest of the country is Greek Orthodox), cows, and a love of cow's-milk cheese.

The island of Tinos, a barren rocky place near the much more famous island of Mykonos, has a few traces of its Venetian past. The majority of its people are Roman Catholic, and between them and the Greek Orthodox, the island, with a population of only ten thousand, is famous for its 575 churches. Several can be found in each rock-bound village, and people come from all over Greece to visit one particular Greek Orthodox church because it is believed to bestow miracles.

Tinos is also known for its Gruyère-style cow's-milk cheese. This traditional cheese that the farmers used to make themselves is a creamy, semisoft young cheese rolled into balls.

The cow breed originally brought to Tinos by the Venetians is long gone, as are the native Greek cows. Now people raise Holsteins. It would be hard to imagine a less suitable place for them. The terrain seems better suited for sheep and goats than for cows, but the sheep and goats are used only for wool.

In the summer, Tinos is a shrubby, brown, mountainous island. Some of the mountains look like piles of rocks. Geologists say that these rocky areas used to be at the bottom of lakebeds. The big round rocks are called *volokos*, which is also the name of the island's most traditional village. Volokos has two-story whitewashed houses with small blue-trimmed windows. The houses are often built against large rocks, and their ground floors are used for storage. The streets are far too narrow to accommodate cars. Most Tinos villages were originally accessible only by donkey.

Wild capers grow in the mountains between the rocks. In the winter the terraced slopes marked off with rock walls are covered with rich green grass. Winter is a good time for the cows here, though it is sometimes a bit harsh, with occasional snowfall.

The island's three dairies collect milk from ten farmers, who have a total of 130 Holsteins. The milk is used for making cheese, as well as yogurt, butter, and ice cream. The leftover whey is shipped to the nearby island of Paros to use in their cosmetic industry.

Greece has fallen on hard times in recent decades, and many farmers have given up working the land and moved to Athens. In fact, all over the Greek islands, people are giving up and moving to Athens. Life is hard here. There is no doctor and no airport, only ferries that run three times a day to Mykonos and other nearby islands. Those farmers who remain support themselves by producing a little cheese, growing tomatoes and potatoes, and making *raki*, a clear 90 percent smooth alcohol distilled from grapes and anise.

Near the main Tinos port and an unattractive sprawl of modern white buildings, Adonis Zagradanis raises thirty cows on 3,000 square meters of valley and steeply sloped land. Like all the farmers here, he also has a second job—he owns a butcher shop in town. He grows corn in the dirt-floored valley to feed his cows and also buys leftover corn from breweries in Athens and Thessaloniki. In the winter, he drives his herd up to terraced fields to graze on grass, supplementing the grass with corn and feed.

Despite the landscape and summer heat, the milk produced by cows here has a high fat content, especially after the spring grass has come in, making it ideal for cheesemaking. The farmers have milking machines available, but prefer to milk by hand because they say that the machines cause mastitis, an infection of the teats. But this usually happens only if the machines are not properly cleaned.

Ionnis Armaos, another Tinos native, lives here, but his wife and young son and daughter live in Athens, where she works. They get together on weekends. Ionnis toiled as a construction worker for a while, but the trade pulled him away from his native island and so he decided to buy a dairy and make Tinos cheese. To do this, he hired an experienced cheesemaker, Polykarpos Delatolas, to teach him how to make the traditional cheeses and a few others of his own invention.

Ionnis's San Lorenzo dairy is located high in the mountains, where in the hot summer the temperature is almost 4 degrees centigrade cooler than in most of the rest of the island. In a spotless two-room dairy he receives 5,000 to 6,000 kilos of milk a day, out of which he makes 50 kilos of ten different varieties of cheese. His is a successful operation. He struggles to keep up with demand. Much of his product is sold retail in his little shop.

His little dairy survives, the Holsteins survive, and Tinos remains a cheesemaking place.

SOME BELIEVE THAT yogurt promotes longevity. And that is key to the success of Icelandic skyr, which is really a fresh cheese but became popular as a yogurt-type food.

For thousands of years, the sour taste of yogurt was greatly appreciated in some countries, including Greece, India, Turkey, Bulgaria, Iran, and the nations of the Arab world. But many deemed it too harsh and sour for Western palates.

Then, in the beginning of the twentieth century, Élie Metchnikoff, a Ukrainian Jew who was deputy director of the Pasteur Institute, became the first to study probiotics, the beneficial qualities of bacteria and yeast. He also had a theory about immunity that earned him the 1908 Nobel Prize for medicine. He noticed that Bulgarian peasants sometimes lived past the age of 100 and that yogurt was central to their diet. He concluded that the live cultures in yogurt had properties that ate away diseases and slowed down aging.

And that is how France became the first yogurt-eating Western nation. During the course of the twentieth century, many other Western nations followed its example, so that by the end of the century yogurt was a very common food in the West.

But there was a problem: Westerners didn't really like yogurt. They did indeed find it harsh and sour, and they didn't like the way the whey seeped out between spoonfuls. So during the twentieth century, Western dairies started to change yogurt. First, they added sugar to mask the sourness. Next came sugary fruit compotes. Then low-fat dieting became fashionable, and skim milk and powdered milk began being used. And in 1970 frozen yogurt was invented. By this point there were no longer any live cultures in the Western yogurt and the Western yogurt eater probably wasn't going to live as long as a Bulgarian peasant. As the food writer Anne Mendelson wrote, "Yogurt had been turned into a kind of premixed sweet-and-sour pudding or pseudo-ice cream."

But there were still some Westerners who wanted it plain, nonfat, and healthful, and they discovered Icelandic skyr, which the Icelanders and some Scandinavians have been enjoying since the time of the Vikings. It is a traditional food made from the milk of traditional Viking cows that graze on hearty Icelandic grass.

Aside from the switch in Viking times from sheep to cows, there has not been much change to skyr over the centuries. It looks like yogurt but is denser and creamier. Since it is a cheese, it is also more expensive and time-consuming to make than yogurt. One liter of milk makes one liter of yogurt, but it takes three liters of milk to make one liter of skyr.

Skyr is sold in the same type of plastic containers as yogurt and naturally contains zero fat. But like yogurt, skyr leaks whey. Realizing that this is an unpopular characteristic, the largest skyr maker, MS Iceland Dairies, developed something called ultrafiltration, which keeps the whey from leaking—an ironic historical twist, because originally skyr was made to produce whey.

This new skyr has a silky creaminess and a higher protein content. Sales greatly increased, and the product also spread to America. At first, its makers were frustrated because the Americans thought it was yogurt and sold it in the yogurt section. The label clearly says skyr, not yogurt, but if Americans want to buy their cheese as zero-fat high-protein yogurt, why not? After all, zero-fat yogurt was very popular in the United States, while skyr was unknown.

BRITISH CHEESE, LIKE most British artisanal products, was hit hard by industrialization in the second half of the nineteenth and first half of the twentieth centuries. One by one, many great and famous cheeses disappeared. But since then, farmers have been bringing them back, and now the British claim to make more different types of cheese than the French. This is a somewhat dubious claim, probably based on Charles De Gaulle's famous question, "How can you govern a country which has 246 varieties of cheese?" As was his habit, the general was understating the problem—there were probably more than 246 varieties at the time. Today, France lists about 450 types of cheese, and within those types are numerous varieties. Modern experts estimate that France makes about 1,000 kinds of cheese and Britain 700, though among the British offerings are imitations of such cheeses as Brie, now made in both Cornwall and Somerset.

The British list of cheeses was probably longer in the eighteenth century, but it is much longer today than it was in 1950. Among the cheeses on the list now are Duddleswell, created in Sussex in 1988 and made from sheep's milk and vegetable rennet; Lincolnshire Poacher, made from unpasteurized cow's milk since 1992; and Stinking Bishop cheese, for which the cheesemaker Charles Martell of Gloucestershire revived a nearly extinct cow breed, Gloucester cattle. The milk for Stinking Bishop cheese is pasteurized, and when there is not enough milk from the Gloucester cows available, milk from neighboring Holsteins is added. The cheese has been mentioned

Milk Street, London, mentioned as far back as the twelfth century for selling milk
(near Honey Street and Bread Street). This illustration by William Blake shows
May Day, 1784. Milkmaids in garlands and chimney sweeps are dancing down
Milk Street. (HIP/Art Resource, NY)

in many films and television shows, adding to its popularity, as film-
makers can't seem to resist its name. But actually, the cheese is
named not for its smell but for the "perry," a local pear cider, in which
it is immersed during production.

Among the extinct cheeses brought back to life in Britain is the
Wiltshire Loaf, one of the most famous cheeses of the eighteenth
century. It disappeared because of its alleged similarity to cheddar,
which started to be produced inexpensively in various factories around
England in the nineteenth century. Wiltshire could not compete.

In Saxon times, the area west of London known as North Wiltshire was used for grazing sheep, but in the thirteenth century, as
more forest was cleared, it became pastureland for cows. Dairying
was women's work, but it was widely believed—and this idea
endured for centuries—that for a woman to make cheese, she had

to be married. Unmarried old maids would be bitter and their milk would sour.

The word "dairymaid" came from the old English word *dheigh*, which originally meant a female dough kneader, a breadmaker, and head of household. After the Norman Conquest, when cheesemaking moved out of the house and into a special room, that room was called a *dheigh* and then a "deyhouse." By the twelfth century, cheesemakers in Wiltshire were called "dairymaids."

By the seventeenth century, the valley of the River Avon, a soft north-to-south curve through the upper west corner of Wiltshire known as the Wiltshire Vale, was a dairy area known for its cheese. Pasturelands were tidily divided by dark-leafed hedgerows, and the average dairy had about a dozen cows, a small number compared to the herds of dairy farmers in competing cow-cheese regions such as Cheshire.

Longhorns, the first English-bred cattle, were a beef breed whose milk did not work well for cheesemaking. They were crossbred with Dutch and Scottish breeds to make the shorthorn, a smaller cow, and this was the breed that was favored in Wiltshire. Even when good cheese cows such as Ayrshires, Jerseys, and Guernseys came along, they were not popular in Wiltshire because local cheesemakers found their cheese to be too rich, or, as they expressed it, "too buttery." In fact the Wiltshire farmers preferred their cows to graze in poorer-quality pastures so that their milk would not be too rich.

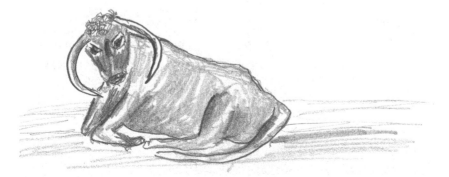

English longhorn

Shorthorn

In the late eighteenth century, a small, ten-pound Wiltshire cheese was called a truckle, and an even smaller version, famed throughout England, was known as the Wiltshire Loaf.

By the late nineteenth century, however, cheesemaking was vanishing. It was difficult to find dairymaids or any kind of dairy workers anymore, and the Wiltshire farmers were having trouble competing with the industrialized cheese plants. It seemed most people just wanted cheddar. Then in the 1930s, the British government started guaranteeing milk prices that made selling milk more profitable than selling cheese. And that was the end of Wiltshire cheese.

At least it was the end until Chad and Ceri Cryer started making it again in 2006. Chad did not grow up on a farm, and when he was a child, he had a toy cow with six teats. He didn't learn that cows had only four teats until he became a biologist. He was an expert on bees. While teaching, he met Ceri, another teacher, who had grown up on a Wiltshire farm. Her father, Joe Collingborn, an outspoken Wiltshire farmer, was the third generation on a dairy farm his family had owned since 1910. The farm was in the green, tall-grass valley of the Avon, a picturesque countryside with clay soil and fields that were ill-suited for anything but dairy. The Collingborns have been in Wiltshire since at least 1086; they are in the 1086 Domesday Book, which lists English landowners.

Weary of teaching, Ceri wanted to return to her family farm, and Chad agreed to the move. He thought it would be a chance for him to raise bees for honey. But instead he learned to be a dairy farmer and learned that cows did in fact have only four teats. The couple replaced the family's English shorthorn stock with a type of Holsteins that are not quite as large or productive as the American Holstein, but also not as expensive to feed. "Not as high an output, but not as high an input," Joe said. They bred them to have strong legs and strong feet and to keep producing for about ten years. They had one that was still producing at fourteen years. They maintain about ninety cows.

The Cryers produce 2,000 liters of raw milk a day and sell most of it to a cooperative. But they keep 5 to 6 percent for themselves and pasteurize it. With this they make yogurt flavored with gooseberry, strawberry, raspberry, and rhubarb. They also sell thick cream, ice cream, nonhomogenized milk, and four kinds of cheese. Their farm, which they call Brinkworth Dairy, produces the once famous Wiltshire Loaf and three other cheeses of their own invention: a strong, flavorful blue called Brinkworth blue; a softer, milder blue called Royal Bassett blue; and a soft, fresh garlic-and-pepper cheese.

Chad spends weekends selling at outdoor farmers markets in various towns, including London. But cheese, for all its proud tradition, is not how the Brinkworth Dairy makes money. "The most profitable product," says Chad, "is ice cream, because it's all air, and it's a highly competitive market. Cones are most profitable, but it has to be a hot sunny day." In any event, cheesemaking is back in the Wiltshire Vale.

JANE AUSTEN, THE great chronicler of the English gentry in the late eighteenth century, has her characters Emma and Mr. Elton in *Emma* eating Wiltshire and Stilton cheese. They may have been the most popular upmarket cheeses of the time. The British have always loved blue cheese. That is why two of the four cheeses at Brinkworth Dairy are blue.

There are well over thirty Great Britain blue cheeses, some of them new. In Scotland's Clyde Valley, Dunsyre Blue, a mild, complex blue cheese made from unpasteurized Ayrshire milk, was invented in 1980. Beenleigh Blue, which also dates from 1980, is made from vegetarian rennet and pasteurized sheep cheese. But the most famous British blue cheese is Stilton. No one knows its exact origin, but it began to be widely traded in the early eighteenth century.

In 1990 the European Union defined the standards that cheese must meet in order to be called by the name Stilton. Oddly, it defined it as what Stilton was at that time, not what it was historically. So a cheese that is made in the town of Stilton, which is where the cheese got its name, does not have the right to call itself Stilton. Rather, the rules specify that the cheese must be made in one of three counties: Derbyshire, Leicestershire, or Nottinghamshire. Stilton lies outside these counties. The rules also say that Stilton must be made with pasteurized cow's milk. But the original Stilton was made long before Pasteur and so was made with raw milk. Using pasteurized milk to make Stilton only began in 1988 when a batch of the cheese was falsely accused of being contaminated, causing all the Stilton makers to panic and switch to pasteurized milk.

"Stilton," The Illustrated London News, *November 4, 1876.*

In 2004, Joe Schneider, a new cheesemaker, decided to make authentic Stilton that would resemble the cheese sold in 1730. This is called "farmstead" cheese, and all the milk used to make it is raw and from the same herd. Raw milk changes day to day and from season to season, making for subtle and sometimes more obvious variations in the cheese. The newer Stilton cheese, in contrast, is made from milk from a variety of herds, and after that milk has been pasteurized, there is no variation in the cheese. Its producers proudly call this consistency. But all those rogue flavors that turn up in Stilton made from raw milk are dimensions of taste that are simply missing from pasteurized cheese.

In Britain, a cheesemonger can be fined for selling Schneider's authentic old-style Stilton as Stilton or "raw milk Stilton." And so the traditional Stilton has ended up with a name as odd as its situation. It is called Stichelton. If you want authentic old-time Stilton such as Joe Schneider's, you have to ask for Stichelton.

The worst nightmare, the kind of scare that drove Stilton cheesemakers to pasteurization in 1988, occurred again in the summer of 2016. There was an outbreak of *E. coli*, and a three-year-old girl died. The Food Standards Scotland traced the bacteria to a small dairy called Errington Cheese, the makers of Dunsyre Blue. Their stocks were seized and the twelve-person workforce laid off. Other tests by respected laboratories such as Actalia, a French company, also tested the cheese but saw no evidence of *E. coli*. The microbiologist in charge of the test, Ronan Calvez, said, "a great deal of cheese consumed in France is made of unpasteurized milk. The laboratories that test the safety of these products are of the very highest microbiological standards, and have developed sophisticated testing regimes to ensure that no bacteria containing harmful toxins are present in any cheese sold."

It is the certified-milk-versus-pasteurized-milk argument all over again. One of the reasons why cheese was made in the first place is that, long before Pasteur, people knew that cheese, in contrast to milk, did not make you sick, not even in hot countries. Most cheeses are made with a salt that kills bacteria, and blue cheeses often contain penicillin. But countries that fought long and hard for pasteurization, such as the United States and Britain, are still sensitive on the subject. Ironically, France, the land of Pasteur, isn't. Historically, France has been much more open to the use of raw milk than either Britain or the United States. In fact, many French cheesemakers insist on raw milk, which is probably why the French are the world's greatest cheesemakers, even if their thousand varieties make them hard to govern.

19

THE SEARCH FOR CONSENSUS

THE OLD DEBATES over such subjects as which animals have the best milk, whether it is better to breastfeed or artificially feed, what is the most humane way to care for milking animals, what is the best food for cattle, and whether milk is a healthful food for adults have not been resolved. Newer arguments over such subjects as raw milk versus pasteurized milk have not been settled either. And now there are even newer disputes—about hormones and organic farming and genetically modified organisms.

The farmer usually respects the land and respects his animals, but faced with the narrowest of profit margins, his foremost question is often not so much "What is right?" as "What works economically?" With milk prices held down by the federal government in the United States, the European Union in Europe, supermarket chains in Australia, and various other governing bodies all over the world, held to barely above the cost of production, farmers must look for ways to make their milk and milk products stand out so that they can demand higher prices. The hormone-free milk, organic farming, and GMO-free movements all represent such opportunities, but they also make milk more expensive to produce.

The challenge for farmers is to find the right equation, including the right size herd and the right animals, to make their dairies economically viable. Some products, such as the best cheeses, are very expensive to produce, but they also have a market—there are people willing to pay for these prestigious luxury items. There are far fewer people willing to buy expensive milk. There is too much cheap milk available.

IT IS CLEAR that the issue of raw milk has not been resolved. But there is unusually broad agreement among cheesemakers and cheese lovers that raw milk is needed to make good cheese. Not everyone agrees, but there is far more consensus on this issue than there is on the issue of liquid milk for drinking. In America, where almost all milk is pasteurized, many people are not even aware of the debate. The popular belief is that it was a real concern in the nineteenth century, but then pasteurization was adopted and now people don't get sick anymore. But the debate has never ended.

It was never said that certified raw milk is unhealthful. In fact, there are a lot of reasonable arguments positing that it is more healthful than pasteurized milk. The argument has always been that pasteurized milk is easier to monitor and regulate than certified milk. But even in countries where raw milk is legal and regulated, such as France, it is very hard to find. In fact, European supermarkets specialize in unrefrigerated shelves full of UHT milk. It is really the same argument: The milk isn't as good but is much easier to regulate and handle. Good milk in Paris is usually found at quality *fromageries*, where with luck you might even find it unhomogenized.

Raw milk is currently benefiting from a growing interest in "all natural" food. Raw milk certainly is that, and it is also noncorporate, since no corporations are interested in producing it. In the United States, twenty-eight states allow raw milk under certain conditions, and there are strong movements in other states to loosen laws against it.

In the twenty-first century, more than half a million Americans a year buy raw milk, and that number is slowly rising. In Canada, where laws make it difficult to purchase raw milk, consumers buy shares of a herd so that they are not buying raw milk, but simply drinking what is theirs.

Raw milk is an example of the kind of trade-off farmers must calculate into their operations. Raw milk is a small market and it would probably be hard to sell the amount produced by a large herd. But there is a market, and a farmer might be able to make a profit if he produced just the right amount because raw milk advocates are willing to pay twice the normal milk price for it.

Some claim that raw milk prevents allergies and other diseases. Scientists generally discount these claims, a position that some raw milk advocates see as a part of a conspiracy. That theory is supported by the fact that the U.S. Food and Drug Administration, which has always been suspected of being too cozy with Big Milk, has gone so far as to warn that raw milk contains dangerous pathogens and should not be consumed.

What is going on? If the government thinks raw milk is dangerous, how can the states allow it? Because lawmakers and consumer advocates in many states believe that the government is being too sweeping in their pronouncements in the name of simplicity and that under the right circumstances, raw milk can be safe.

In the state of New York, raw milk is legal, but it can only be sold on farms. The risk from haphazard distributors or careless stores has been eliminated. People wishing to buy raw milk have to find a farm and visit it, which can be problematic for many New York City dwellers, one of the largest markets for allegedly healthful milk specialties.

Mary and Bill Koch have five acres north of New York City in the Hudson Valley, an area that is part suburban and part agricultural. The Kochs decided that a better way to keep their grass trimmed would be to get a cow. "If you have grassland," said Mary, "and you don't use it, you wind up spending a lot of time and money mowing it." Also, she added, Teddy, their then second-grade son, became upset when he found that the mower was killing bird eggs, rabbits, and birds—"all kinds of wildlife chopped up."

Once the Kochs got a cow, however, they realized that they knew nothing about keeping cows. The cow died. It takes a great deal of knowledge to keep a cow healthy. Undeterred, they studied up on cows and got a few more. Today, they generally maintain between three and five Dutch Belted cows, handsome black cows with a thick white stripe around the middle. Despite their name, they are thought to have originated in Switzerland or Austria. Not as large, hungry, or productive as Holsteins, they are an affordable and reasonably productive breed. They produce good butterfat and, as Mary said, will "eat almost anything and turn it to milk."

The Kochs mostly let their cows graze, but they supplement their food with hay, especially in the winter. They also feed them treats such as carrots and Swiss chard from their garden. "Cows love getting treats," Mary said. But even grass-feeding isn't as cheap as it sounds; it costs eight dollars a day to feed each cow.

The Kochs tried to earn a living by selling their dairy products. But they couldn't earn enough to survive and so turned their farm into a four-room bed-and-breakfast, which they named Thyme in the Country B&B. They try to offer their guests a "true farm experience" by serving them raw milk along with other farmhouse products such as yogurt, butter, eggs, chicken, and pork. People often drive up from New York City to have "a farm experience" at Thyme in the Country, where they can enjoy raw milk. All the milk is raw, and inspectors come to the farm regularly. The Kochs find them to be likeable, and the inspectors have found the Kochs extremely helpful.

Dutch Belted cows

The Thyme's habitués seem to be passionate about raw milk. Upon a new guest's arrival, they often rush up to him or her to grill them about their attitudes toward raw milk, to know whether the guest is for it or against it. It doesn't seem to occur to them that some people might be completely neutral on this issue.

In contrast, in the state of Idaho, raw milk can be sold in a store. It commands a high price because it is expensive to produce. A separate herd has to be maintained and regularly tested.

Alan Reed in Idaho Falls thought there might be a good market for this high-priced raw milk and decided to give producing it a try. He sold his raw milk for $6.87 a gallon as opposed to the $5 usually charged for a gallon of regular whole milk. It was a luxury item for milk gourmets: nonhomogenized milk with the cream still floating on top. But he only sold about 130 gallons a week. "I was surprised," he said. "Everybody was coming in here talking about raw milk. I thought I would sell more."

THE DEBATE OVER breastfeeding versus artificial feeding will probably never end. History has shown it to be a constant, with the majority opinion over which is better regularly swinging back and forth, like the fashionable width of neckties. In the early twenty-first century, especially in the United States, the pendulum has swung back to breastfeeding. Once regarded as the choice of poorer people who could not afford to buy milk or formula, it is now regarded as the thing to do by upper-middle-class and wealthy women. Working-class women, on the other hand, especially women with jobs, now prefer bottle-feeding.

Starting in 1920, scientists, doctors, and mothers began doubting the virtues of breast milk. It had been found that the vitamin content of breast milk varied among women, and that not all women produced good breast milk. Artificial feeding, in contrast, could be consistently nutritious, and gradually, it became more popular. There were regional differences. Midwesterners preferred breastfeeding and

Northeasterners preferred artificial feeding. But in general, artificial feeding was increasingly popular. During World War II, doctors assumed that there would be a resurgence in breastfeeding because there was a scarcity of milk. But that never happened.

Hospitals pushed the notion of "scientific" childcare and discouraged breastfeeding. This was highly significant because by 1950, 80 percent of American women gave birth in hospitals, up from only 20 percent in 1920.

Historians argue over whether breastfeeding became popular because of the seven Catholic housewives who in 1956 founded the La Leche League or the league just happened to come along at the right moment. Both theories might be right.

La Leche began when its founders, Marian Tompson and Mary White, were breastfeeding their babies at a Christian Family Movement picnic and decided that they should do something to promote what they termed "natural mothering." This struck a chord with American women, because it was a reaction against "scientific mothering." There was a strong feeling that doctors should not be telling mothers what to do.

Since most doctors at the time were men, the La Leche movement had a vague feminist overtone, even though the nascent feminist movement itself was generally on the side of bottle-feeding. But in other ways, the La Leche movement was decidedly nonfeminist. White had eleven children, and she and her league believed that women should stay home, have many children, and dedicate all their time to childcare. They openly opposed the idea of women working.

After the Christian Family picnic, White and Thomson met with five other women in White's Franklin Park, Illinois, home. Between them they had given birth to fifty-six babies, and their sentiment was that they were the ones with some real expertise.

Five years later, there were forty-three La Leche groups around the United States. In 1976, twenty years after their founding, there were about three thousand groups. But breastfeeding actually reached its lowest point in popularity in the 1970s. In 1971 only 24 percent of American women breastfed and only 5 percent continued to do so after six months. The best food for babies, most people were convinced, was commercially bottled baby formula.

NOTHING CHANGES POPULAR thinking as quickly as a major scandal.

Until 1970 few statistics were available on infant mortality in the developing world. But when researchers started gathering these statistics, they discovered horrors reminiscent of the epidemics that had swept through nineteenth-century cities. In India in 1970, 2.6 million babies died. Worldwide, an estimated 11 million babies every year did not live to their first birthday. Most of the deaths were in Southeast Asia, the Indian subcontinent, and Africa.

In these countries, breastfeeding was on the decline and formula feeding was on the rise. It was a by-product of the struggle to develop the economy. Women were going to work. Parents were leaving their villages for work in the cities. Children were being left in the hands of relatives or paid caregivers, and without mothers being present, the practice of feeding babies formula became widespread. Companies that made formula, especially Nestlé, Bristol-Myers, Abbott Laboratories, and American Home Products, saw a tremendous opportunity and began building plants and producing formula throughout the developing world.

Baby formula became a huge industry. By 1981 the world formula market was estimated to be worth $2 billion, with Nestlé controlling over half of it. Much of Nestlé's growth was in poor countries, where people responded to extensive advertising campaigns on the health benefits of formula. Formula was also given out in hospitals, because market research showed that the majority of mothers given formula in hospitals would continue using it after they went home. Later investigations revealed that some of the "nurses" who visited mothers in hospital maternity wards were actually Nestlé's salespeople in disguise. The United Nations World Health Organization estimated that between India, Nigeria, Ethiopia, and the Philippines, bottles were being given out at a rate of five million each year.

Then health experts working in these countries found a notable jump in infant mortality as well as in the number of cases of gastro-enteritis and malnutrition. This was clearly connected to the use of formula.

It was not that the formula was toxic or lacked nutrition. It was just ill-suited for the developing world. It came in a powder form, to be mixed with water. Most poor women had little access to clean water. They could boil water for the requisite twenty minutes, but this involved burning some kind of fuel that they typically could not afford. Maybe just a few minutes would help? It wouldn't. Also, while the women knew the importance of washing the bottles between uses, the water with which they washed them usually wasn't clean and often infected the bottles. Many babies died from germs lurking in washed bottles.

A woman left the hospital with a small supply of free formula. By the time it was used up, she was no longer lactating, so the decision whether to breastfeed was now out of her hands. When she tried to buy more formula, she was surprised to discover that it was incredibly expensive, a significant part of the family income. So she stretched the formula by mixing it with water. Some babies were getting mostly water, and it was often tainted water.

What next happened was similar to the public campaign against swill milk in the nineteenth century. In 1973 a British magazine, *New Internationalist*, published an exposé, "The Baby Food Tragedy," that broke the story. Other articles followed, as did lawsuits, a public outcry, a movement to boycott the formula companies, and a 1978 U.S. Senate investigation led by Senator Edward Kennedy. The United Nations then established worldwide regulations to monitor advertising campaigns.

The baby formula companies survived, but they had lost the public trust. And it wasn't just that mothers no longer trusted formula companies. They no longer trusted formula itself. Between 1971 and

1980, the percentage of American mothers breastfeeding beyond six months rose from 5 percent to 25 percent. International organizations that had encouraged bottle-feeding, such as UNICEF, started to advocate breastfeeding.

By the second decade of the twenty-first century, 79 percent of American mothers were breastfeeding their children for the first six months, and 49 percent of American mothers were continuing to breastfeed beyond six months. In New York City, 90 percent of mothers were breastfeeding; mothers who never breastfed were becoming a rarity.

And tremendous claims were being made for breastfeeding: It would reduce the risk of ear infections, pneumonia, a sometimes fatal intestinal infection called necrotizing enterocolitis, sudden infant death syndrome (SIDS), allergies, eczema, asthma, type 2 diabetes, leukemia, cardiovascular disease, behavioral disorders, and even low intelligence. Unfortunately, medical research has ruled out most of these claims completely and found little evidence to support the rest.

One of the stronger arguments for breastfeeding is what is known as "attachment theory." This contends that a happy, healthy child needs to establish a strong bond with its parents at an early age. Breastfeeding, it is claimed, creates a strong mother-child bond.

The great irony is that while the move toward breastfeeding was in part an attempt to escape the duplicity of profit-hungry large corporations, other corporations have discovered that, in America at least, there are tremendous profits to be made from breastfeeding. The U.S. government's push for breastfeeding has also become a push for breast pumping. This new progressive approach allows babies to benefit from the nutritive value of breast milk while freeing mothers from the burden of breastfeeding. A mother can now pump her breasts and save her milk in bottles, so that while she is at work, a caregiver can feed her baby. So the first thing to go was the attachment theory.

When Bill Clinton was president, the White House had a breast-pumping station for staff. Barack Obama's Affordable Care Act agreed to cover the purchase of breast pumps. Private insurers followed. And once health insurance covered breast pumps, sales exploded. It has become a huge business in the United States: the industry expects breast pump sales in 2020 to reach $1 billion, and related paraphernalia sales are expected to reach another $2 billion.

This means that human milk has become a very valuable commodity in the United States. There is now an oversupply of breast milk because breast pumps induce mothers to produce more milk than their babies need. (Milking machines for cows do the same thing.) Women who pump often find their freezers full of unused milk. Then the question becomes, what to do with it?

There is a century-old tradition of milk banks. First established in Boston in 1910, their mission has traditionally been to provide one mother's excess breast milk to another mother who needs it. But today, some customers, believing that breast milk has medicinal purposes, are buying it when they are ill. Some athletes, who believe that it will make them stronger, are also buying it. Other customers sell soap made from breast milk, and when a London ice cream shop started selling vanilla-with-lemon-zest breast-milk ice cream for more than twenty dollars a scoop, they could not keep up with the demand.

Breast milk can now be bought or sold online at websites such as www.onlythebreast.com, where many dubious health claims for human milk are made. The sellers are free to say whatever they wish. One, reported on National Public Radio, stated, "My milk makes giants." But like other types of milk, human milk from sources that are not well regulated are problematic and can be dangerous. In 2015, doctors at Nationwide Children's Hospital in Columbus, Ohio, tested 102 samples that they ordered online and found that 10 percent had cow's milk added.

According to the researcher Sarah Keim, 13,000 women were buying and selling breast milk online in 2011, and by 2015, that number had risen to 55,000. She also found that 75 percent of the milk that she bought online had bacterial contamination and disease-causing pathogens.

HUMANS ASIDE, THE issue of what milk is the best milk is also still far from settled. Although cows dominate throughout the world, that has been a pragmatic decision. It has not been agreed that cow's milk is superior, only that cows are the most efficient milkers, although some Asians also see the advantage of buffalos. Outside of India, few think cows produce the best milk, except, of course, cow's milk producers.

There have always been, and still are, passionate goat's milk fans. Donkey's milk has also retained a following, especially since it has been determined that people who are allergic to other milks have no problem with donkey's milk. But donkeys do not produce large quantities of milk. The Swiss donkey's milk company Eurolactis has one thousand donkeys but struggles to produce enough milk. Donkey's milk is also in demand for cosmetics, and Eurolactis has started a line of milk chocolate bars that they say are lighter and more nutritious than normal milk chocolate bars because donkey's milk has less fat and is rich in omega 3 fatty acids.

Camel's milk, which is very similar to cow's milk, is still produced in a few places, notably Mauritania. Australia is also trying to establish a camel's-milk industry. In the nineteenth century, the British imported camels to the Australian outback for transportation and construction projects. But soon trucks and jeeps arrived, and the camels, being of no more use, were released into the wild. Since then, their population has grown and may now number a million. Dairy companies are now attempting to harvest their milk.

Asians, especially in the Philippines and Southeast Asia, still greatly value buffalo's milk but in recent years have been turning to cows. In Vietnam, the Afimilk company built twelve dairies with 32,000 cows. They began importing Holsteins from New Zealand in 2010 and growing feed. Their operation is now too big for grazing. The cows are fed in feedlots, where they are lined up in front of food trenches.

An even bigger dairy in an even less likely place is a Saudi Arabian farm with 180,000 Holsteins. Fidel Castro's idea of keeping cows in air-conditioned barns is in use here and doesn't seem as crazy in this wealthy desert land as it did in Cuba. Here, it is only a question of how great a financial loss is acceptable. It helps to be an oil-producing kingdom that provides cheap energy. The dairy, the pet project of Deputy Crown Prince Mohammad bin Salman, is supposed to be an example of Saudi Arabia building up industries other than oil. But keeping that many cows comfortable, and milk chilled and shipped around the Arabian Peninsula that is the world's second-largest desert, takes energy. So do the nine thousand refrigerated vehicles that distribute the milk.

The worldwide tendency for milk to be produced in ever larger farms has raised awareness about animal welfare issues. First, there is the issue of space. While it is true that farms with larger herds have more acreage, the acreage per head tends to decline. Few cows have as much space as the three on the Kochs' little five-acre farm in the Hudson Valley. Second, there is the issue of food. Thomas Jefferson in his farming notes wrote, "The number of cattle to be kept on a farm must be proportioned to the food furnished by a farm." But today farms rarely grow all of the food that their cows need. In some cases, they provide none. In the Saudi Arabian dairy, all the feed is flown in from Argentina. Patrick Holden, who runs an organic dairy farm in Wales with a herd of Ayrshires from which he makes cheddar cheese, said, "Most farms are like airports, with fertilizer, feeds, everything coming in from all over the world." His goal, which he has not entirely achieved yet, is to have an entirely sustainable farm, a farm that supplies all of its own needs.

There is also another issue. Cows are sensitive creatures. No one who has ever looked into their big soft eyes would doubt that. Goats are mischievous, sheep are insecure, and cows are sensitive and sweet.

It has been understood for centuries that happy cows produce more milk. Cows that are stressed produce poorly. In 1869 Catherine Beecher advised, "Gentle but firm treatment make a cow easy to milk and to handle in every way." And in 1867, Annabella Hill wrote, "No animal repays kind and generous treatment more than the cow."

A number of studies suggest that music makes cows happy, leading some dairies to provide music for their cows. It is widely believed that cows have a preference for classical music, but there is no science to support this. At Hawthorne Valley Farm in Ghent, New York, musicians come every Christmas to sing carols to the cows. If some Americans find this odd, Indians would see nothing strange about it.

Many farmers give names to their cows because they are convinced that cows enjoy being called by name. Sometimes they just choose names they like and sometimes they choose names that are memory aids, to help them remember which cows are which, though the cows usually have numbered tags as well. At the Brinkworth Dairy in Wiltshire, England, they name their cows to distinguish their blood-lines. Each cow of the same bloodline has the same first initial— Candy, Ceri, Cookie. The cows also have numbers, and Chad knows a cow's name only after he sees her number, but his father-in-law, Joe, knows every cow's name as soon as he sees her.

The Ooms family started their upstate New York farm in 1950, but they come from a long line of farmers who have owned dairy farms in Holland since 1500. They have four hundred cows, which is considered a large dairy for that part of New York, and though they have a reputation as large-scale, no-nonsense farmers, they name each of their cows. But four hundred is a lot of names, and when asked for a few examples, Eric Ooms turned shy and said, "Oh, I'm terrible with names." But he did recall that New York senator Chuck Schumer had visited the farm and so there was one cow named Chuck and a second named Schumer.

Alan Reed has a small farm in Idaho that was started by his uncle in 1955. The farm was in the country at first, but since then Idaho Falls has grown and now the dairy with between 140 and 250 Holsteins is in the suburbs. A tall, lean, quiet Westerner, Reed sneered a little when asked if he named his cows. "We don't have time to name them," he said. But then he admitted that he had had to change that after children started visiting his farm for regular events and were very upset to learn that the cows did not have names. So he instituted naming contests, and now many of the cows do have names. Another Idaho farmer, with forty-four hundred cows, Jordan Funk, said, "I only name them if they step on my toe."

Basque dairy farmer Tambourin said that he names sheep that display particular personalities. "But," he said, raising his index finger didactically, "never name a pig." He declined to explain why.

But whether farmers name their animals or not, they eventually have to kill them.

The lifespan of a cow depends on how well she is treated. In some of the huge dairies where thousands are lined up in feedlots, cows only live three or four years. Smaller, gentler farms can keep them a few years longer, even to the age of ten. A cow could even live well into her teens, some to twenty or even older, if not overworked. A cow's natural lifespan is thought to be about twenty years, and one was famously recorded to live to be forty-nine.

But older cows do not keep lactating. A cow eats more than 60 pounds of feed every day, and on most farms, feed is 70 percent or more of operating costs. It can cost $70 or more a week to keep a cow. That means that a modestly sized farm with only one hundred cows would spend $7,000 a week to feed its herd. This is one of the reasons why it is so difficult to make a profit from milk. Each cow has to produce more in the value of her milk than she costs to feed, and when she stops producing, the farmer can no longer afford to keep her. This is true in the big "factory" farms as well as the small family farms.

When a cow stops lactating, a truck usually arrives to take her away to a slaughterhouse, where she becomes hamburger. Cows are the leading source of hamburger meat in America. Dairy cows are very lean because they put everything into producing milk, so they make good lean ground meat.

Lorraine Lewandrowski, a lawyer who also owns a dairy farm in the Mohawk Valley, New York (increasingly, farmers survive by having two careers), once had cows named Aviva, Anneke, Benazir, Celeste, Esther, and Fiona. Each cow had a name with the first initial of the cow's mother, and Lewandrowski talks about their "sweet and trusting eyes." And yet she eventually has to kill them. She finds it particularly sad to have her cows end their lives with a long, frightening truck ride, followed by death alone in a strange place. So she will often have them shot on the farm. As Patrick Holden said, "Dairy farming is tough."

Another tough reality of dairy farming is that a newborn calf is usually separated from its mother within a few hours and bottle-fed. Farmers have always found this to be a key to having a successful dairy. In the frieze of "the dairy of al-Ubaid," five thousand years ago, the heifers wear muzzles to keep them from suckling their mothers.

The mothers mourn their loss with loud moans and big sad eyes, sometimes for days. There are no eyes sadder than those of a sad cow. But not all cows are natural mothers. According to Ronny Osofsky of Ronnybrook, a popular dairy north of New York City, "Some cows are very maternal and moo a lot. Others don't care. Some cows feel maternal toward every calf they see."

Sometimes cows are in visible emotional pain following the loss of their calves, but most farmers accept this as part of dairy farming. If a calf was left to suckle on the mother for a few months, as nature intended, the cow would be happier and the calf healthier, but the farmer might lose what small profit he had. When the young grow up to depend on the farmer, not the mother, it makes the farm more manageable.

According to Patrick Holden, the separated calves are upset for two or three days and the mothers are sad for a week. He said that some farmers have been experimenting with letting the cow suckle for three months, but he pointed out that a cow has a calf a year—occasionally twins—and if you let it suckle during the first three months, you lose a third of the milk for that year. "There is a lot that is upsetting about dairy farming," he said.

IN THE LAST two decades of the twentieth century, more and more multi-thousand-cow dairies started up opening up while many small family dairies went out of business. Large-scale industrial farming was the key to survival. The most traditional dairy regions, New England and New York, were the biggest losers as the dairy industry moved west, to states where there was space for giant industrial farms owned by corporations, not families. California became the largest milk-producing state in the country. The state is now popularly known for its anti-industrial food movements, but those movements were born out of the fact that California is the heartland for industrialized food. The second-largest state for milk producing is Idaho, traditionally a farming, lumber, and mining state that is new to the dairy industry.

Even though the consumption of milk had started declining in the United States in the twentieth century, there were new uses for it. Beginning in the 1920s and 1930s, the dairy industry started using organic chemistry to transform elements of milk, whey, and buttermilk into industrial products. The field was called chemurgy, and the first milk-derived products were made out of the protein casein, found in a high percentage in cow's milk, which as early as 1900 had been used for such things as making buttons and a coating for airplane wings. Casein was also used in glue, in finishing paper for color printing, and for making paint. Skim milk, once the unwanted product left over from making butter, was the primary source of casein, and it took 33 pounds of skim milk to make one pound of casein.

But there were not always enough uses for the extra milk produced. The European Economic Community, the predecessor to the European Union, once paid European farmers to overproduce—though they didn't see it that way. They simply agreed to buy whatever milk the farmers couldn't sell. This led the farmers to produce far more milk than could be sold and created a scandal, with the famous European "lakes of milk" and "mountains of butter" going to waste. Europe then, to be less wasteful, switched to the "single farm payment," with a single payment given once a year to every farm based on its acreage and compliance with environmental practices.

In America there is no similar subsidy for milk, but farmers are simply spurred on to produce more by the economic principle that says the more milk produced, the more profit to cover expenses.

In the United States, as in all countries with a dairy tradition, people once regarded dairy farms fondly. Milk industry advertising captured perfectly the beloved image of a dairy farm that most Americans had—a red barn with white trim, hilly green grass country, and sweet-faced brown Jersey cows. But this scene hardly exists any more. Today dairies are muddy, sprawling affairs with long feed lots, milking parlors, and other prefabricated structures.

For those who can afford it, the milking parlor of choice is a rotary milker—based on the same idea as the Rotolactor shown at the 1939 World's Fair, only larger and computer-regulated, with cows eagerly waiting to step on. It is still the same comical bovine merry-go-round with the cows lining up for their ride. In most milking parlors, cows grow restless, stamp their feet, and defecate, but this is not so on their rotating trip, which they thoroughly enjoy. When they get back to the starting point and have to step off, they clearly show their disappointment at leaving.

People do not want to live near a dairy anymore. Cows defecate and they are also extremely flatulent. This was never an issue with the charming forty-cow farm with the little red barn. But when a few thousand cows live next door, farting and producing mountains of manure that the farm endeavors to dry out and convert to fertil-izer, they are very strong-smelling neighbors.

Farmers with large herds usually have more manure than their pastureland can absorb. The total animal waste in the United States is one hundred times more than what is taken in by human sewage treatment plants, and 4.5 million people are exposed to dangerous nitrate levels in drinking water, mostly from poorly handled animal waste.

The odors of large dairy farms contain chemicals and gases that can cause respiratory and digestive ailments. They also contribute to climate change. A United Nations study concluded that cattle flat-ulence produced more greenhouse gases than do automobiles. The big culprit is methane gas, which, although it does not get as much attention as carbon dioxide, is more than twenty times as destruc-tive in terms of climate change. A dairy cow burps and farts between 300 and 400 pounds of methane gas every day. And that figure does not include the roughly equal amount that emanates from her manure. Standards and procedures have been established to deal with these issues, but farmers say that the procedures are very costly— one more expense digging into their already very narrow profit margin. For example, farmers in some areas are no longer permitted to hose down their barns and stables. Instead, they have to cover piles of manure with straw to absorb the animal waste and then use that straw as compost. This is a very sound and careful way to deal with waste, but it involves increased manpower and costs considerably more than does simply hosing.

THE PUBLIC THAT no longer loves dairy farms is also questioning dairy practices, with some arguing that cows today are producing unnatural quantities of milk. It certainly is unnatural in that they are producing ten to twenty times more milk than they would need to suckle their young. Of course, dairy farming has always been based on the premise that cows need to overproduce if a farm is to survive, but the amount of that overproduction has never been as high as it is now. Cows aren't living as long as they once did, either, and they have more health problems. The huge udders needed to hold all that extra milk create back and leg problems.

Animal rights activists also claim that dairies often abuse their animals by leaving them tethered to the same spot all their lives without a roof over their heads. But this happens only rarely. Dairy farmers want to get the most out of their cows and understand that if they treat them well, they will produce more milk. It has even been shown that cows that eat in covered feedlots produce more milk than cows that eat in lots that don't have roofs.

Many activists and consumers today are also pushing for dairy cattle to be grass-fed. The irony is that grass-feeding—simply letting cows eat the grass that grows in pastures—is the absolute cheapest way to raise cattle.

England has a good climate for pastureland, and Chad Cryer of Wiltshire's Brinkworth Farm said that the grass there is good for grazing ten months of the year. But, he added, "We are not impressed with grass-feeding. Cows need more." At Brinkworth, the Cryers give their cows more by spending money on high-protein grains such as barley and alfalfa because they believe the cows need the extra nutrition.

Many farmers feel this way about the "grass-fed" idea. No one denies that cows fed in feed lots produce more milk than cows that are grass-fed. That is why they take on the extra expense of purchasing feed.

Eric Ooms of New York State said that grass-feeding would not work for his four-hundred-head dairy herd. "As you get bigger you have to keep track of nutrients," he explained. "If you let cows graze you are not sure how much they are eating. If you stall-feed them you know exactly."

In South Australia the weather is mild enough for grazing, but during the summers, December to April, there is little water and the land becomes parched. The dairy farmers grass-feed their cows as long as they can and then turn to store-bought feed. At Nang-kita Hills, the Connor family, four-generation dairy farmers, said that they would grass-feed their cows year round if they could, because they would like the savings of not buying feed. The Connor farm spreads over hilltops with strong sea breezes—the southern coast of Australia, only twenty miles away, faces Antarctica—and experience has taught them to be cautious of investment. They do not disapprove of feed-lot dairies, but say, "If you invest in feed lot and don't get high production you lose a lot. Grazing is less risky."

In South Australia, whether they feed their cows or graze them, farmers have one unique problem: Dozens of strange creatures as large as human beings with powerful hind legs and wimpy little fore-arms hop onto their land and eat all their cows' food. Some farmers will shoot one or two kangaroos to scare off the rest, which scatter in a mad and bouncy retreat.

Kym and Kate Bartlett live on a dairy farm that Kym's family has owned since 1927. They have a year-round garden and elegant palms and other decorative plants around their farmhouse. Their land is irrigated by the wide Murray, Australia's longest river, and they are careful not to let the water from their farm run back and pollute the river. They grass-feed their two hundred Holsteins year round. "You have to move the cows every day," said Kate, "but the cheapest way to produce milk is to get them to eat grass."

Australian Illawara

Karen and David Altmann's Dakara Farms is also on the Murray, but they had difficulty preventing their farm water from running back into the river, and so they decided to operate a feedlot dairy. Those who are against feedlot dairies often describe them in dark and denouncing tones, but Dakara Farms seems clean and well managed and the cows well cared for. The Altmanns have four hundred Holsteins and one hundred Illawarra, an Australian red cow produced by crossing shorthorn and Ayrshires with local breeds. The Illawarra are high producers, and their milk, rich in butterfat, is valued for butter- and cheesemaking.

David, a fifth-generation dairy farmer, bought this farm in 1999 and built an open-air feed shed with a roof and long row of metal bars with spaces for cows to stick their heads through and eat the food piled up on other side. It is a cool place in the heat of summer,

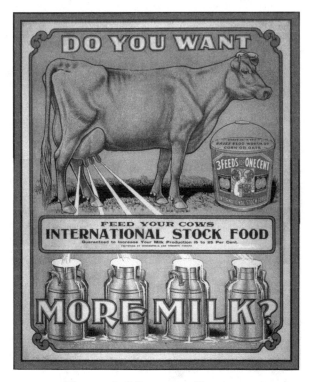

1920s ad for cow feed "guaranteed" to increase cow's milk production.

with mist sprinklers that are turned on in extra-hot weather. The feed smells fragrant—a mixture of hay, wheat, canola, apple pulp from a nearby juice producer, crushed oranges from an orange juice plant, potatoes, and leftover grain from brewers.

The cows are free to wander the hills and drift over to the feedlot when they are hungry. They produce well over the average cow's yield in Australia, and their lives must not be too stressful, because they lactate for at least seven years, often nine, and sometimes even twelve or fourteen. But eventually, like all cows, they get shipped off to the slaughterhouse.

20

RISKY INITIALIZATIONS

WITH EVERY SCANDAL, like the nineteenth century one about swill milk, the one involving Nestlé formula, or the more recent one in China, the consumer grows more distrustful of dairies, corporations, and government. There are always more scandals, and so public distrust of milk only increases.

The nuclear age began in 1945 with the U.S. bombing of Hiroshima and Nagasaki. Instead of horrifying the world, these bombings set off a nuclear arms race. While the Japanese were struggling with the aftereffects of radiation and fallout, the U.S. government denied that such things existed and pursued nuclear tests aboveground, sending poison into the atmosphere. In 1949 the Soviet Union exploded its first atomic bomb. Britain followed in 1952, by which time the United States was exploding even more powerful hydrogen bombs in the South Pacific. In 1953 the Soviet Union was testing hydrogen bombs in Siberia. Scientists estimate that between 1945 and 1958, the power of the nuclear weapons exploded in the world was equal to eight hundred times the power of the Hiroshima bomb.

With each explosion, tiny particles that could not be seen, not even with a high-powered microscope, entered the atmosphere and traveled the world. Some of this so-called radioactive fallout dissipated, but some didn't, including strontium-90, which accumulates in bones, especially the unformed bones of children, and can cause cancer, leukemia, or premature aging; and iodine-131, which accumulates in thyroid glands and can also cause cancer. This fallout landed on the plants that cows ate, and it passed on into their milk and into anyone who drank that milk. In 1958 a test of milk in forty-eight U.S. and Canadian cities showed that its strontium-90 content had at least doubled between 1957 and mid-1958. The U.S. government tried to assure everyone that those levels weren't dangerous, but few people, including many scientists, believed them. In 1962, health officials found so much iodine-131 in Salt Lake City milk that they advised people to avoid it.

In 1963, the United States, the Soviet Union, and Britain agreed to stop aboveground nuclear testings in the atmosphere, but nuclear latecomers France and China did not sign on; France continued until 1974 and China until 1980. Since then testing has been underground. But in 1963, the Federal Radiation Council warned that even without further explosions in the atmosphere, health issues such as an increased incidence of leukemia and birth defects could continue for the next seventy-five years. Contaminated children could pass on those conditions to their children as well.

IN THE 1950S there was a widespread movement to avoid milk because cows were eating nuclear-contaminated grass and their milk contained the contaminants. As this risk appeared to abate, other problems arose, and milk suffered from a series of scandals. Each came with initials: PBB, rBGH, BST, BSE, and GMO.

In 1973 in Michigan, cows were becoming lethargic. They had stopped eating and were producing less milk. Eventually, some became too weak to stand. But veterinarians could not identify a disease. Perhaps pesticides were the cause? Farmers suspected a new protein-enriched feed that contained Nutrimaster, a brand name for magnesium oxide, which increased milk production by improving digestion. After considerable testing, it was discovered that the feed contained not magnesium oxide but polybrominated byphenyl, or

PBB, a flame retardant for fabrics and plastic. When consumed by cows, the PBB was stored in their livers and fatty tissue. From the fatty tissue, it went into milk, and from the milk, into the fatty tissue and livers of people who drank the milk, causing severe health problems. Selling milk from cows so sick they could barely stand? It seemed as if not much had changed since the days of swill milk.

It turned out that the Michigan Chemical Corporation, which made both Nutrimaster and PBB, had mixed the two products up because they looked similar and were packaged similarly. The U.S. Department of Agriculture assured the public that there was nothing more to worry about, and going forward, that was true. But a few years later, the children who had consumed the contaminated milk started to show alarming symptoms such as weight loss, hair loss, lack of coordination, and lethargy. Those children who had breastfed from women who had consumed the contaminated milk didn't escape, either; the mothers had had PBB in their breast milk. A third of all Michigan farm children were sick by the mid-1970s, along with some children in Detroit and other cities. After the PBB scandal, the public grew increasingly wary of high-intensity feeds, which increase milk production but often cause indigestion in the cows.

Starting in 1993, cows began being injected with something called recombinant bovine growth hormone, or rBGH, also sometimes known as bovine somatotropin, or bST, which producers promised would increase a cow's milk production by as much as 25 percent. To facilitate that increased production, the cows' udders grew to an unnatural size. That was troubling to some dairy farmers, but others reasoned that if a farm could increase production that much without buying and maintaining additional cows or even buying additional feed, it was worthwhile.

The rBGH compound had been developed by the Cornell University College of Agriculture and Life Sciences, which was also exploring other ideas for large-scale, high-production dairy farming. Although the college was located in upstate New York, most of its ideas were not suitable for the small New York or New England dairy farms, but rather for the large dairy farms in the West. Many there, including California farms with as many as thirty thousand cows, began following the Cornell model and using rBGH.

There has since been a significant consumer outcry against rBGH, in large part because it was created using highly controversial genetic modification methods, with the help of an old consumer-target favorite, the Monsanto Agrochemical Company. In 1993, it was approved for use by the Food and Drug Administration, but Canada, the European Union, and other countries banned the use of rBGH in cows, fearing that consuming milk from cows injected with the hormone heightened the risk of certain types of cancer. There is no conclusive evidence that such a risk exists, however, and in the United States, the National Institutes of Health concluded that milk from rBGH-injected cows and milk from noninjected cows were exactly the same. The American Cancer Society refuses to take a position on the issue but says that so far there is no evidence of risk.

Nonetheless, rBGH has not been a huge success in the United States. Some farmers who started using it in the 1990s have discontinued the injections, and today, less than a third of American farmers use the hormone. Farmers have found that rBGH has not lived up to expectations as far as milk production is concerned, and that it causes cows to suffer indigestion and teat infections.

This in turn has caused farmers using the hormone to use more antibiotics, which is another concern for milk drinkers. It has been found that excessive use of antibiotics in farm animals leads to resistance to these drugs in humans who consume these animals' meat or milk. Most research does not indicate that milk from antibiotic-injected cows poses a risk to humans, but a growing number of Americans no longer respond to antibiotics, and about 23,000 Americans die from antibiotic-resistant infections every year. There is also a risk that the antibiotics given to the animals could get into the soil and into plants and vegetables, resulting in even more humans' becoming antibiotic-resistant. In 2013, in an action unrelated to the use of rBGH, the FDA decided to clamp down on what they regarded as the excessive use of antibiotics in cows, chickens, and pigs.

Despite the government's and scientists' assurances that rBGH is not harmful, many American consumers have been demanding that labels say whether the milk, butter, cheese, or yogurt they buy has been made with milk from hormone-injected cows. Some dairy companies have already started to do this, selling milk in containers labeled "rBGH-free" or "hormone-free."

At the heart of the controversy is the fact that these large-production giant dairies that can fill all the milk shelves of America at a low price are putting small farms out of business, and developments such as rBGH encourage that process. In fact, the Cornell economists studying the impact of rBGH predicted that it would drive small farms out of business because of the type of herd management it requires. It is central to the Cornell philosophy that "larger herds are indicative of better managers." They even encouraged economic policies to help ease small-scale farmers out of dairying, which would completely undo the rural culture, even the landscape, of a number of states such as New York and Vermont. Vermont senator Bernie Sanders wrote in 2012, "There is something very wrong when large processors reap large profits, and family farmers . . . can barely survive, or must sell their farms."

Between 1970 and 2006, about 573,000 dairy farms in the United States closed, but there was not a corresponding drop in U.S. milk production. Why? In the twenty-first century, the number of dairy farms with more than two thousand cows doubled.

CONSUMERS' DISTRUST OF the dairy industry can be blamed not only on PBB and rBGH, but also on BSE, or bovine spongiform encephalopathy, more commonly referred to as mad cow disease. This, too, was caused by an attempt to give cows a concentrated feed that would cause them to produce more milk. The more protein cows eat, at least in theory, the more milk they will produce. And so the idea was hatched that cheap meat and bone meal could be put in their feed, despite the fact that cows are herbivores by nature and not designed to eat meat.

BSE is thought to have first shown up in English cattle in 1985, although the disease was not diagnosed until 1986. It seems to have been caused by the presence of infected nerve tissue—brains and spinal cords—in the feed. The disease, which kills cows, is also fatal and incurable in humans. The human variant of the disease is known as Creutzfeldt-Jakob disease (vCJD).

A major BSE outbreak began in Britain in the late 1980s and early 1990s, with cows becoming more aggressive, having difficulty walking and standing up, and producing less milk. It lingers in the minds of the British public how the British government was slow and modest in its response. The Conservative government then in power derived many votes from rural farm communities and so was reluctant to take draconian measures and order the slaughter of cows, which would have been devastating to farmers. In the end, however, many more cattle were slaughtered than would have been necessary if the government had responded more quickly.

At first, the British government insisted that the disease presented no threat to humans. The grim-faced agriculture minister, John Gummer, who later became a prominent environmentalist, publicly fed his daughter a hamburger in 1990 to show there was no risk. But then house cats and zoo animals began dying from eating food made with beef by-products, and in 1995, the first human victim died of vCJD.

Finally, in the mid-1990s, the British government prohibited the feeding of meat and bone meal to cows and banned the export of British beef for three and a half years. The British health secretary also publicly admitted for the first time that the disease could be transmitted to humans, and 4.5 million cattle were destroyed. By 2015, the UK human death toll from the disease stood at 177 people, and worldwide, another 56 have died of vCJD, including 4 in the United States. The government also at first failed to compensate farmers for giving up their infected animals for slaughter, and as a result, they were understandably reluctant to do so. Even when the government did start compensating them, the payment was considerably less than they could earn by selling the animals as food. And since initially the government was insisting that there was no threat to humans, the farmers felt no moral obligation to take the loss.

Today, the mad cow disease outbreak and its mishandling are well remembered, and that may be one of the reasons why the British public leads the world in the global fight against genetically modified organisms (GMOs). Despite an almost complete lack of scientific proof that GMOs are harmful, the cause has become a popular one.

From the milk perspective, the central GMO issue is that American dairy cows are fed GMO grain, which is grown in the United States. Strangely, the question is if it is all right for genetically modified cows to eat genetically modified food. GMO grain is banned in many other countries, including those belonging to the European Union, although there could be an economic motive behind this, since the banned grain is grown in the United States and if European farmers were to use it they would be buying imported American grain instead of the European product.

In 1999, at a celebration of the first twenty-five years of biotechnology, James Dewey Watson, a member of the three-person team that won the 1962 Nobel Prize in Physiology or Medicine for revealing the structure of DNA, the double helix, said that the field of GMOs offered "certain promise and uncertain risk." Watson's point of view was that to make progress, you had to take risks.

GMOs are indeed a field full of promise. They could produce food that no one would be allergic to, or food that when eaten prevents tooth decay, even perhaps food that fights certain diseases. Since farmed fish draw sea lice, they could produce lice-rejecting fish. It almost seems that for any problem, there could be a genetic solution.

Some of the claims of those protesting the use of GMOs appear to have a measure of validity, but one accusation that is easily refuted is that GMOs are forced on farmers against their will. On the contrary, U.S. dairy farmers talk about how GMO grains, less expensive than non-GMO grains as well as insect-resistant, have literally saved the struggling family farm. And today, as the GMO companies are consolidating, farmers are expressing concerns that fewer companies will lead to fewer laboratories and the invention of fewer products. Farmers want GMO products much more than consumers do. In fact, Monsanto, the leading GMO developer, concedes that they made a mistake at the outset of marketing their products by focusing on winning over farmers, not the general public.

In the early decades of GMO development, scientists discussed the risks, such as new pathogens and the possible evolution of mutant viruses, and believed that experimentation in the field had to be controlled. By the early 1970s, the idea emerged, perhaps never a good one, that the scientists themselves would regulate the process and, when necessary, put the brakes on experimentation. But who were the scientists assigned to the task? Many were the employees of large multinational corporations, so that in the end, the corporations were regulating themselves.

While there is a lack of scientific evidence to support the claims that GMOs are physically harmful, one sociopolitical claim is irrefutable: If the plans of Monsanto and other GMO companies were to be realized, those few companies would completely control the production of agriculture throughout the world. The plants grown from GMO seeds do not produce GMO seeds, so the only way to maintain GMO crops is to keep going back to the companies and buying more seeds—something that was not foreseen when the GMO crops were first developed. Curiously, the anti-GMO lobby is far more vociferous in denouncing this sociopolitical fact than the farmers themselves, who are the most directly affected.

Farmers often work from a different economic reality. Most dairy farmers depend on growing or purchasing high-protein grains such as alfalfa and corn to feed their cattle. In the United States, GMO grains dominate this market, and dairy farmers do not want to be forced into buying expensive "GMO free" grain.

The dispersion of GMO crops, like all crops, is hard to control. In 2000, one GMO corn, StarLink, designed to feed cattle but unfit for humans because it is highly allergenic, unexpectedly started turning up in human corn products. At first it was found only in taco shells, but later it appeared in popcorn and other products, some of which were exported to other countries. StarLink had been planted in only a very small percentage of cornfields, but still got into much of the U.S. corn supply.

Another GMO crop problem concerns weeds. Early on, Monsanto created crops that were resistant to weeds, eliminating the need for toxic and expensive herbicides. But in time, it was found that the weeds could develop a tolerance for these GMO crops, creating unstoppable superweeds.

Insect-resistant GMO crops also raise troubling issues, proving James Watson right: There is great promise in GMO crops, but also certain risks. What if a crop keeps away not only unwanted insects, but also wanted ones? Alfalfa reproduces through the pollination of bees. Without bees, there would be no alfalfa. Any other means of alfalfa pollination would be prohibitively expensive. But bees are mysteriously disappearing. Is this attributable to the GMO crops? Scientists have been intensely studying this possibility, but so far they have no certain answers.

Alfalfa is an important issue for dairy farmers, as it is a leading high-protein grain for milk-producing cows. Alfalfa cattle feed is a $4.7 billion annual business. Why did Idaho, a state with little dairy history, become the nation's second-largest milk producer? Because irrigation projects created huge expanses of arable land, and much of it was planted with alfalfa. This was possible only because in the 1950s, bee experts learned how to transport alkali bees, a Western species that is particularly adept at pollinating the somewhat tricky alfalfa flower, to Idaho in large numbers. But now that these bees, like the honeybees and other bee species, are disappearing, what happens to the Idaho dairy industry and much of the Western dairy industry?

There are many theories as to why bees are disappearing, ranging from the proliferation of cellphones to the use of pesticides to the development of GMO crops. A few of these theories, such as the cellphone one, have been disproved, but most have not. However, the problem with the GMO theory is that bees are also disappearing in countries that have banned GMO crops.

If GMO products are as safe as Monsanto says they are, the GMO companies have made a grave error in not submitting their products to rigorous regulation. And the U.S. government is making the same mistake in not enforcing stricter regulation. GMO scientists have failed to prove that GMO products are safe, and GMO opponents, scientists included, have failed to convincingly demonstrate that they are not. This leaves a distrustful public with no one to believe. When Qiao Yanping, a longtime bureaucrat in the Chinese state dairy industry, was asked how he felt about GMOs, he answered, "I don't like to use GMOs. I refuse to import GMO alfalfa. They may not be dangerous, but there is not enough science to assure their safety. Safety now is the big issue." And many in other countries feel the same way.

The *New York Times* conducted a study of GMOs that was published in October 2016. Relying on academic research, independent data, and industry research, the study was most notable for what it didn't find. It didn't find scientific support for the claims that GMOs were harmful. By this point that was not surprising news. What was surprising was that the study found no advantage to using GMOs. The American and Canadian GMO crops compared to the European non-GMO crops showed no increase in yield and no reduction in the use of pesticides, and the use of herbicides had actually increased. After twenty years of hard fighting pro- and con-GMOs, nothing has been gained and nothing lost.

THE GMO BATTLE has created an opportunity, however. Small dairies are always looking for ways to make their milk special, because if it is special, they can charge more for it—and survive. Dairies producing special milk, such as milk from cows fed non-GMO crops, are increasingly turning up near New York, Chicago, San Francisco, Los Angeles, and other cities. Cities have well-educated, affluent people who are willing, even eager, to pay more money for special milk. Patrick Holden, the zealous organic farmer in Wales, said, "The city dwellers are going to lead the revolution."

But most farmers are not trying to lead a revolution, they are just trying to make a profit. Ronnybrook in upstate New York successfully sells hormone-free milk in New York City for almost double the standard milk price. Its owner, Ronny Osofsky, is also trying to become GMO-free. "It's not easy," he said. "Ninety percent of soy and eighty percent of corn is GMO." Asked if he thought the use of GMOs was harmful, he said, "No. I don't think it's harmful. I don't want to use it because of the public reaction to it." He allows no hormones for the same reason, although he is not convinced that they are harmful either. He just does not want to use what his customers dislike.

Osofsky has found that his expensive products—milk, yogurt, cheese, and ice cream—are popular in New York City. His milk is special not just because it is hormone free, but also because it comes in glass bottles. "It is more expensive to sell milk in glass bottles but it tastes better," he said. And it makes it look like a prestigious product—not just another carton of milk.

Osofsky prides himself on being good to his herd of one hundred Holsteins of a particularly large strain with names such as M&M and Talia. "I treat them gently," he said. "Cows are like dogs. If you are nice to them, they are nice to you." And here is an important point: A farmer with a hundred cows can treat them like pets but a farmer with a thousand cannot. When his cows are not out grazing, they are in barns with black-rubber-covered foam rubber mattresses on which to relax. He feeds them mostly grass, but also some grain. He would like to claim that they are entirely grass-fed, but finds that they need a little extra protein to be productive.

Still, the pressure is relentless. "The truth about dairy prices," he said, "is that if the price is low you have to sell a lot, and if the price is high you have to sell a lot while the price is high."

He has considered becoming organic, but he finds it cruel to refuse antibiotics to a cow with an infection. Once a cow has been treated with antibiotics, her milk can no longer be sold as organic, even long after the drugs are out of her system. Large farms with thousands of cows can establish separate nonorganic herds, but if a farm has only fifty or one hundred cows, it cannot afford to do that. Many other farmers also say that they find the rules of organic farming to be too cruel to cows. Even more say that those rules make it too expensive for a small farm. The *New York Times* once described Ronnybrook as "beyond organic."

In 2002, Dan Gibson, a former New York City business executive, decided he wanted to start a different kind of dairy farm. He had bought a thousand-acre farm in the Hudson Valley, and although at first he allowed its occupant to continue running their dairy, he soon decided he wanted a change. "I realized I hated the dairy business, kicking calves to the curb to get milk," he said. He started thinking he could turn his dairy into an "animal-friendly farm." The milk would have to be more expensive, but people in New York City would be willing to pay more because his product would be better and they would like the way he did things. "People want this really badly," he said. Then he explained, "I learned in marketing that to sell something you have to make it different, better and special. I produce pure Jersey, grass-fed, Animal Welfare Approved milk."

The American Welfare Approved (AWA) label, started in 2006, is intended to assure consumers that the meat and dairy products carrying their label are produced on farms that are kind to animals. The animals must be grazed and grass-fed, and the farmer must follow practices that are not harmful to the environment. An AWA-approved farm cannot be organic, however, because one of the requirements is that sick animals get antibiotics when needed.

On Dan Gibson's farm, newborn calves stay with their mothers for months. The farm's emphasis is on quality, and there is no concern about quantity. Eric Ooms, a non-AWA New York farmer, can get as much milk from one of his large Holsteins as Gibson can with his entire herd of fifty small brown Jerseys. One half-gallon glass bottle of Gibson's milk sells for $7, but even after two years of full production and consistent selling at high prices, his farm has yet to break even, let alone show a profit. Not many farmers can afford to do that.

Alan Reed in Idaho Falls once thought of going organic. But he did not have enough land. It takes a great deal of land to grow organic feed and meet the extensive organic grazing requirements. But, he added, "I try to buy organic produce for my family."

Cory Upson of the Belted Rose Farm in the hill country near Cooperstown, New York, had to completely change his approach to dairy farming in order to survive. In 1998 he had fifty-five Holsteins in a conventional dairy producing grade-A milk at the low minimum price that the government had set—under $10 for a hundredweight, 11.6 gallons, at the time. Then he became an organic dairy farmer. Asked why he switched, his answer was straightforward: "We didn't make any money." He had had mostly Holsteins, but had noticed that his two Dutch Belted cows prospered without the grains that his Holsteins seemed to need. So he gradually switched to a herd of twenty-three Dutch Belted cows, which are entirely grass-fed on his hills, and became entirely organic.

"To make more money," he explained, "you increase revenue or reduce expenses." He radically reduced his operating costs by becoming an organic farmer. He no longer buys grain and is now training horses to replace tractors to reduce his fuel expenses. He has fewer than half as many cows as he had before, and his cows are producing less than half as much milk. But organic milk is priced on the assumption that people will pay more for it, and he is selling his milk at a high price to Horizon, the largest organic milk producer in America. "I'm not getting rich, but I am making money now," he said.

When Americans buy organic milk, they imagine it coming from a small family farm, but almost all organic milk in America is produced by large corporations like the one to which Cory Upson sells. In other countries, some small-scale organic farms exist, such as Holden's Welsh farm, but they are rare. In Wiltshire, England, there are some small, eighty-to-one-hundred-cow, organic farms. Their production is low. Chad Cryer said, "You could not feed the world organically."

In South Australia, where organic milk sells for twice the price of regular milk, it is still a struggle to make it profitable. Denise Richie of the Hindmarsh Dairy wanted her goat's milk to be certified organic. "They asked for wild and wacky things," she complained. She complied with one wild thing after another, seeing her costs go up and up. Finally she had to give up when they demanded that she replace all of her fencing because the wood had been treated with a chemical.

In central Idaho's Magic Valley, farmers have made organic milk work for them. In this region, where irrigation brought alfalfa and alfalfa brought dairy, there is little tradition of small family dairies. It is an area of large farms, including some companies that moved in from California. Here, the Funk family owns a large stretch of flatland in the valley, with jagged, snow-capped mountains called the South Hills on the horizon. Jordan Funk said, "To Easterners they are mountains. To us they are hills." The Funks have been family farmers here for four generations, and their story is the story of farming in the area. They started with potatoes, a mainstay of Idaho agriculture. Then they switched to sugar beets. But in 1997, when milk was booming in Idaho, they decided to start an organic dairy.

From the start, they understood that becoming organic entailed a great deal of work and expense. Double Eagle Farm is no longer a family operation. They decided that the best way to defray costs was to manage a large herd of 4,400 cows on 8,500 acres of arable land with seventy-five to one hundred workers. Once they made that decision, they were forced to operate more like a company. Jordan Funk said, "I told my dad, if we are going to be organic, we have to be big enough to hire a secretary for all the paperwork." Among other things, organic farming requires a lot more paperwork than most farmers are used to.

To become certified as an organic farm is a long and complex process. It takes three years of using no pesticides, herbicides, or chemical fertilizers just to get the ground certified. Then the feed has to be entirely organic. Organic feed is extremely expensive to buy, and also costly and difficult to grow. No weed-resistant GMOs are allowed, so the Funks have to regularly weed their fields.

If a cow gets sick, the Funks treat her with antibiotics and then sell her in auction, sometimes to another dairy and sometimes as meat. A few cows get sick every week, but with 4,400 cows, they can afford to give up a few. They pasteurize milk and feed it to young calves. Feeding calves milk is another added expense of organic farming. In regular dairies, calves are fed a formula, which is much less expensive than pure milk. But in organic dairies, pure milk is required.

The trick of organic farming is to try to keep cows from getting infections, something that is all the more difficult to achieve when there are thousands of cows. Dirk Reitsma at the Sunrise Organic Dairy near the Funks' farm said, "Organic farming is all about prevention." He uses an Israeli-invented system. Each cow has a thermometer on its leg and there are "labs" in the milking stalls that monitor the health of the cows and warn the farmers if there is any sign of sickness. "If you catch it early, we can just give some vitamins and don't need antibiotics," Reitsma said.

Organic rules require a minimum of 120 days of grazing a year, and during that time, 30 percent of the dry matter eaten must be from the farm's own pasture. The Funks grow about half the hay they use, and almost all the barley, but they have to buy very expensive non-GMO proteins—canola, soybeans, and flax.

The Funks' milk is sold raw to Horizon for double or even triple the regular milk price. And that price is guaranteed in three- or five-year contracts, so it is very stable. This is one of the advantage of selling to a large company. For most dairy farmers, the instability of milk prices is one of the greatest challenges they must deal with; it makes planning very difficult.

ORGANIC FARMING STARTED in the 1960s, but organic milk became popular only in the 1990s. Revelations in 1991 about antibiotic residue in nonorganic milk made some people switch to organic. People wanted to know that their milk was produced with special care. Once organic milk became widely available, it became the top-selling organic food. Organic milk took on a number of burning issues. The cows were not to be treated with hormones or antibiotics, or fed GMO grains. There was also a general belief that organic cows were better treated than cows in crowded factory dairies.

With milk, popular perception is more important than science. The issue is whether customers will accept a higher price for milk that is made in a special way. There was no exact definition of what organic was until the U.S. Department of Agriculture established rules in 1997. In 1998, when organic milk was new, sales were $60 million in the United States. But unfortunately, meeting the government requirements for organic farming was prohibitively expensive for small farms. Today, Horizon alone, the largest organic milk producer, sells $500 million worth of organic milk a year. It is one of three large companies that have cornered more than 90 percent of the organic milk market.

Horizon buys milk from six hundred organic farms around the country, some of which, such as the Upsons' Belted Rose, are quite small. But all the milk from both large and small farms gets mixed together in tanks and packaged as Horizon. Large national companies were not what the enthusiasts of the organic food movement had in mind. The organic movement is tied to the locavore philosophy—the belief that quality food comes from small local farms that produce with great care because they know their customers. This is why dairies like Ronnybrook that are not organic are preferred by some city customers to organic dairies.

BEYOND ALL THE modern debates about hormones, antibiotics, genetic modification, and chemicals is a fundamental question that after ten thousand years has still not been answered. That is, if a dairy did everything right and its milk was perfect, would it be good for you?

After all, adults drinking milk is not natural. Neither, for that matter, is babies drinking anything but their mothers' milk. The 60 percent of the world that is lactose-intolerant are made the way nature intended humans to be.

But then there is also what the biologist E. O. Wilson called "the natural fallacy" to consider. The fallacy, according to Wilson, is the belief that whatever is natural is always what is best. To take medicine, wear clothes, or read is unnatural. And to farm, to grow food rather than gather it from the wild, is unnatural.

An organization that calls itself the Physicians Committee for Responsible Medicine urges a dairy-free diet and denounces feeding animal milk to children. When the National Dairy Council, which can rally herds of physicians to back up their claim that milk is healthful, ran a "Got Milk?" campaign, the Physicians Committee retaliated with a mocking "Got Beer?" campaign.

There is considerable evidence that consuming large quantities of whole milk can raise cholesterol and lead to heart disease. This is why low-fat and skim milk have become extremely popular. There are also claims that milk can cause certain types of cancer, such as ovarian cancer, but there are as many studies refuting these findings as supporting them. Others claim that milk causes osteoporosis, but that charge seems largely based on the fact that Asians, who drink far less milk, are less affected by this ailment. Asians may also be aided by the fact that they generally get more exercise than do Westerners and eat more vegetables and less protein. It would be interesting to know whether now that milk consumption is on the rise in Asia, osteoporosis is on the rise as well.

Meanwhile, while some are blaming milk for bone loss, many credit it with bone building. Mainstream science and medicine tell us that milk is a leading source of calcium, vitamin D, and other bone-building nutrients, and we are also told that milk helps prevent hypertension.

When the National Dairy Council was formed in 1915 to promote milk, they started telling families that milk would make their children big and strong, which had been the reason the Japanese Emperor had encouraged milk drinking. The National Dairy Council's first brochure was titled "Milk, the Necessary Food for Growth and

Soviet milk maid.

Health." The endorsement of milk by athletes became a tradition, and in the 1960s, the American dairy industry touted the words of Vince Lombardi, the Green Bay Packers coach with one of the longest winning streaks in football history, who said, "I have never had an outstanding successful athlete who was not a hard milk drinker."

There is a persistent belief that milk makes children grow taller, though there is not much scientific evidence behind this. Milk increases something called IGF-1, which has been credited for making people taller, but IGF-1 breaks down in the body and added amounts from milk might have little impact. The entire idea of milk and height may come from the coincidence that children drink the most milk in the years when they are doing the most growing.

TECHNOLOGY HAS COME to dairies, and computers are being used for everything from milking to cow nutrition. Devices like the rotary milker are becoming increasingly automated. Some farmers in England now save wear and tear on the herd with a mobile milk parlor, a machine with stalls that is taken out to the pasture so that the cows do not need to come in for milking. New devices are increasingly worked by robots, and though the technology is currently expensive, many farmers are longing for more robots, because all over the world it is becoming increasingly difficult to find skilled farmworkers. The work is hard, the hours long, and the pay meager.

In the future, there will still be dairies producing milk and all the milk products, and most of the old debates about milk will still be heard. But much of tomorrow's dairy food will be produced by robots. And this will surely open many more new debates on the relative merits or disadvantages of robot-made milk. History has shown that as civilization develops, it creates more, not fewer, arguments about milk.

ACKNOWLEDGMENTS

I never really thought much about milk until Ann Marie Gardner asked me to write something about it for her handsome magazine, *Modern Farmer.* "Later," she said, "you can write a book about it."

I salute the memory of Orri Vigfússon, a kind and dedicated environmentalist, for all his help in Iceland. He had nothing to do with milk, but once he heard that I was writing about it, he started arranging appointments. Orri was like that—an extraordinary man who will be missed.

Thanks to Lorraine Lewandrowski for her help and advice and all the other dairy farmers all over the world who showed me their farms and took the time, a precious commodity for dairy farmers, to talk to me.

Thanks to my great friend Christine Toomey for driving me to the wilds of Wiltshire, and *un merci chaleureux* to my friend Bernard Carrere for taking me to his Basque friends in the heart-stoppingly beautiful Basque Mountains. Thanks to Laura Trombetta for so much help, fun, and adventure, as well as endless yogurt, in China and Tibet.

Thanks to Lata Ganapathy and Rachna Singh Davidar for help in India and to Pankaja Srinivasan for answering all my questions with so many insights.

Thanks to Mirto Siotou for help in Greece, and to Michael Blake from the South Australia government (PIRSA).

Thanks for the help given me by the wizardly Jackson Hole chef Wes Hamilton on a trip through the pass to Idaho Falls. Thanks to Duncan Fuller for his help in central Idaho.

Author feeding a kid on a Vermont farm. (Photo by Dona Ann McAdams)

Thanks to Brad Kessler and Dona Ann McAdams and all the kids for their hospitality and all I learned on their Vermont goat farm.

Thanks to Nancy Miller for editing my book with the same care as the other seventeen she has edited. Thanks to Christiane Bird for her skillful help. Thanks to Charlotte Sheedy for being my friend, adviser, and most wonderful agent.

A special thank-you to the intrepid and bilingual Talia for her assistance in China, India, and Iceland. And to beautiful Marian for all her help.

BIBLIOGRAPHY

Achaya, K. T. *A Historical Dictionary of Indian Food.* New Delhi: Oxford University Press, 1998.

Apple, Rima D. *Mothers and Medicine: A Social History of Infant Feeding, 1890–1950.* Madison: University of Wisconsin Press, 1987.

Ashton, L. G., ed. *Dairy Farming in Australia.* Sydney: Hallsted Press, 1950.

La Association Buhez, eds. *Quand les Bretons passent à Table.* Rennes, Éditions Apogée, 1994.

A. W. *A Book of Cookrye with Serving in of the Table.* London, 1591, Amsterdam: Theatrum Orbis Terrarum, 1976.

Anonymous, edited by a Lady. *The Jewish Manual.* London: T. & W. Boone, 1846.

Artusi, Pellegrino. *La Scienza in Cucina e L'Arte Di Mangiare Bene.* San Casciano, Italy: Sperling & Kupfer Editori, 1991.

Bailey, Kenneth W. *Marketing and Pricing of Milk and Dairy Products in the United States.* Ames: Iowa State University Press, 1997.

Barnes, Donna R., and Peter G. Rose. *Matters of Taste: Food and Drink in Seventeenth-Century Dutch Art and Life.* Syracuse, NY: Syracuse University Press, 2002.

Baron, Robert C., ed. *Thomas Jefferson: The Garden and Farm Books.* Golden, CO: Fulcrum, 1987.

Basu, Pratyusha. *Villages, Women, and the Success of Dairy Cooperatives in India.* Amherst, NY: Cambria Press, 2009.

Battuta, Ibn. Samuel Lee, trans. *The Travels of Ibn Battuta in the Near East, Asia and Africa 1325–1354.* Mineola, NY: Dover, 2004 (reprint of 1829 edition).

Baumslag, Naomi, and Dia L. Michels. *Milk, Money, and Madness: The Culture and Politics of Breastfeeding.* Westport, CT: Bergin & Garvey, 1995.

Beecher, Catherine, and Harriet Beecher Stowe. *The American Woman's Home*. Hartford: Harriet Beecher Stowe Center, 1998 (first edition 1869).

Beeton, Isabella. *Beeton's Book of Household Management*. London: S. O. Beeton, 1861.

————. *Mrs. Beeton's Cookery Book*. London: Ward, Lock and Company, 1890.

Boni, Ada. *Il Talismano Della Felicità*. Rome: Casa editrici Colombo, 1997 (first edition 1928).

Bradley, Alice. *Electric Refrigerator Menus and Recipes: Recipes Especially Prepared for General Electric*. Cleveland: General Electric, 1927.

Brothwell, Don, and Patricia Brothwell. *Food in Antiquity*. Baltimore: Johns Hopkins University Press, 1998.

Burton, David. *The Raj at Table: A Culinary History of the British in India*. London: Faber & Faber, 1993.

Campbell, John R., and Robert T. Marshall. *The Science of Providing Milk for Man*. New York: McGraw-Hill, 1975.

Cato. Andrew Dalby, trans. *De Agricultura*. Devon: Prospect Books, 1998.

Chang, Kwang-chih, ed. *Food in Chinese Culture*. New Haven: Yale University Press, 1977.

Charpentier, Henri. *The Henri Charpentier Cookbook*. Los Angeles: Price/Stern/Sloan, 1945.

———— and Boyden Sparkes. *Those Rich and Great Ones, or Life à la Henri, Being the Memoirs of Henri Charpentier*. London: Voctor Gollancz, 1935.

Chen, Marty, Manoshi Mitra, Geeta Athreya, Anila Dholakia, Preeta Law, and Aruna Rao. *Indian Women: A Study of Their Role in the Dairy Movement*. New Delhi: Shakti Books, 1986.

Child, Lydia Marie. *The American Frugal Housewife*. New York: Samuel S. and William Wood, 1841.

————. *The Family Nurse or Companion of The American Frugal Housewife*. Boston: Charles J. Hendee, 1837.

Cleland, Elizabeth. *New and Easy Method of Cookery*. Edinburgh: Elizabeth Cleland, 1755.

Corson, Juliet. *Meals for the Millions*. New York: NY School of Cookery, 1882.

Coubès, Frédéric. *Histoires Gourmandes*. Paris: Sourtilèges, 2004.

Couderc, Philippe. *Les Plats Qui Ont Fait la France*. Paris: Julliard, 1995.

Crumbine, Samuel J., and James A. Tobey. *The Most Nearly Perfect Food.* Baltimore: Williams & Wilkins, 1930.

Cummings, Claire Hope. *Uncertain Peril: Genetic Engineering and the Future of Seeds.* Boston: Beacon Press, 2008.

Da Silva, Élian, and Dominique Laurens. *Fleurines & Roquefort.* Rodez: Éditions du Rouergue, 1995.

Dalby, Andrew. *Siren Feasts: A History of Food and Gastronomy in Greece.* London: Routledge, 1996.

———— and Sally Grainger. *The Classical Cookbook.* Los Angeles: J. Paul Getty Museum, 1996.

David, Elizabeth. *Harvest of the Cold Months: The Social History of Ice and Ices.* New York: Viking, 1994.

Davidson, Alan. *The Oxford Companion to Food.* Oxford: Oxford University Press, 1999.

De Gouy, Louis P. *The Oyster Book.* New York: Greenberg, 1951.

Diat, Louis. *Cooking à la Ritz.* New York: J.B. Lippincott Company, 1941.

Dods, Margaret. *Cook and Housewife's Manual.* London: Rosters Ltd, 1829.

Dolan, Edward F., Jr. *Pasteur and the Invisible Giants.* New York: Dodd, Mead & Company, 1958.

Drummond, J. C., and Anne Wilbraham. *The Englishman's Food: Five Centuries of English Diet.* London: Pimlico, 1994.

DuPuis, E. Melanie. *Nature's Perfect Food: How Milk Became America's Drink.* New York: New York University Press, 2002.

Dumas, Alexandre. *Mon Dictionnaire de Cuisine.* Paris: Éditions 10/18, 1999.

Edwardes, Michael. *Every Day Life in Early India.* London: B. T. Batsford, 1969.

Ekvall, Robert B. *Fields on the Hoof: Nexus of Tibetan Nomadic Pastoralism.* New York: Holt, Rinehart, and Winston, 1968.

Ellis, William. *The Country Housewife's Family Companion (1750).* Totnes, Devon: Prospect Books, 2000.

Erdman, Henry E. *The Marketing of Whole Milk.* New York: Macmillan, 1921.

Escoffier, Auguste. *Le Guide Culinaire: Aide-Mémoire de Cuisine Pratique.* Paris: Flammarion, 1921.

Estes, Rufus. *Good Things To Eat: As Suggested by Rufus.* Chicago: Rufus Estes, 1911.

Fildes, Valerie. *Breasts, Bottles and Babies: A History of Infant Feeding*. Edinburgh: Edinburgh University Press, 1986.

Farmer, Fannie Merritt. *The Boston Cooking School Cook Book*. Boston: 1896.

Flandrin, Jean-Louis, and Massimo Montanari, eds. *Food: A Culinary History*. New York: Columbia University Press, 1999.

Fussell, G. E. *The English Dairy Farmer 1500–1900*. London: Frank Cass, 1966.

Gelle, Gerry G. *Filipino Cuisine: Recipes from the Islands*. Santa Fe: Red Crane Books, 1997.

Geison, Gerald L. *The Private Science of Louis Pasteur*. Princeton: Princeton University Press, 1995.

Giblin, James Cross. *Milk: The Fight for Purity*. New York: Thomas E. Crowell, 1986.

Giladi, Avner. *Muslim Midwives: The Craft of Birthing in the Premodern Middle East*. Cambridge: Cambridge University Press, 2015.

Glasse, Hannah. *The Art of Cookery Made Plain and Easy Which Far Exceeds Any Things of the Kind Yet Published by a Lady*. London: 1747.

———. *The Compleat Confectioner: or the Whole Art of Confectionary Made Plain and Easy*. Dublin: John Eashaw, 1752.

Golden, Janet. *A Social History of Wet Nursing in America: From Breast to Bottle*. Cambridge: Cambridge University Press, 1996.

Gozzini Giacosa, Ilaria. Anna Herklotz, trans. *A Taste of Ancient Rome*. Chicago: University of Chicago Press, 1992.

Grant, Mark. *Anthimus: De Observatione Ciborum*. Devon, England: Prospect Books, 1996.

———. *Galen: On Food and Diet*. London: Routledge, 2000.

Grigson, Jane. *English Food*. London: Ebury Press, 1974.

Grimod de La Reynière, Alexandre-Laurent. *Almanacs des Gourmands*. Paris: Maradan, 1804.

Guinaudeau, Zette. *Traditional Moroccan Cooking: Recipes from Fez*. London: Serif, 1994 (first published 1958).

Hagen, Ann. *A Second Handbook of Anglo-Saxon Food & Drink: Production and Distribution*. Norfolk, England: Anglo-Saxon Books, 1995.

Hale, Sarah Josepha. *Early American Cookery: The Good Housekeeper*. Mineola, NY: Dover Publications, 1996 (first published 1841).

Harland, Marion. *Common Sense in the Household: A Manual of Practical Housewifery*. New York: Charles Scribner's Sons, 1871.

Hart, Kathleen. *Eating in the Dark*. New York: Pantheon, 2002.

Hartley, Robert M. *Historical, Scientific, and Practical Essay on Milk as an Article of Human Sustenance*. New York: Jonathan Leavitt, 1842.

Harvey, William Clunie, and Harry Hill. *Milk Products*. London: H. K. Lewis, 1948.

Heatter, Maida. *Maida Heatter's Cookies*. New York: Cader Books, 1997.

Heredia, Ruth. *The Amul India Story*. New Delhi: Tata McGraw-Hill Publishing, 1997.

Herter, Christian Archibald. *The Influence of Pasteur on Medical Science*. An address delivered before the Medical School of Johns Hopkins University. New York: Dodd, Mead & Co, 1902.

Hieatt, Constance B., ed. *An Ordinance of Pottage*. London: Prospect Books, 1988.

Hickman, Trevor. *The History of Stilton Cheese*. Gloucestershire: Alan Sutton Publishing, 1995.

Hill, Annabella P. *Mrs. Hill's Southern Practical Cookery and Receipt Book*. New York: Carleton, 1867.

Hooker, Richard J. *The Book of Chowder*. Boston: Harvard Common Press, 1978.

Hope, Annette. *A Caledonian Feast*. London: Grafton Books, 1986.

———. *Londoner's Larder: English Cuisine from Chaucer to the Present*. Edinburgh: Mainstream Publishing, 1990.

Huici Miranda, Ambrosio. *La Cocina Hispano-Magrebí: Durante La Época Almohade*. Gijón, Asturias, Spain: Ediciones Trea, 2005.

Irwin, Florence. *The Cookin' Woman: Irish Country Recipes*. Belfast: Oliver and Boyd, 1949.

Jackson, Tom. *Chilled: How Refrigeration Changed the World and Might Do It Again*. London: Bloomsbury, 2015.

Jaffrey, Madhur. *Madhur Jaffrey's Indian Cookery*. London: BBC, 1982.

Jost, Philippe. *La Gourmandise: Les Chefs-d'oeuvre de la Littérature Gastronomique de L'Antiquité à Nos Jours*. Paris: Le Pré aux Clercs, 1998.

Jung, Courtney. *Lactivism: How Feminists and Fundamentalists, Hippies and Yuppies, and the Physicians and Politicians Made Breastfeeding Big Business and Bad Policy*. New York: Basic Books, 2015.

Kardashian, Kirk. *Milk Money: Cash, Cows, and the Death of the American Dairy Farm*. Durham: University of New Hampshire Press, 2012.

Kelly, Ian. *Cooking for Kings: The Life of Antonin Carême, the First Celebrity Chef*. New York: Walker & Company, 2003.

Kessler, Brad. *Goat Song: A Seasonal Life, A Short History of Herding, and the Art of Making Cheese*. New York: Scribner, 2009.

Kidder, Edward. *Receipts of Pastry and Cookery*. Iowa City: University of Iowa Press, 1993.

Kiple, Kenneth, and Kriemhild Coneè Ornelas, eds. *The Cambridge World History of Food*, vols. 1 and 2. Cambridge: Cambridge University Press, 2000.

Kirby, David. *Animal Factory: The Looming Threat of Industrial Pig, Dairy, and Poultry Farms to Humans and the Environment*. New York: St. Martin's Press, 2010.

Kittow, June. *Favourite Cornish Recipes*. Sevenoaks, England: J. Salmon Ltd., undated.

Kuruvita, Peter. *Serendip: My Sri Lankan Kitchen*. Millers Point, New South Wales: Murdoch Books Australia, 2009.

Kusel-Hédiard, Benita. *Le Carnet de Recettes de Ferdinand Hédiard*. Paris: Le Cherche Midi Éditeur, 1998.

La Falaise, Maxime de. *Seven Centuries of English Cooking*. London: Weidenfeld & Nicolson, 1973.

Latour, Bruno. Alan Sheridan and John Law, trans. *The Pasteurization of France*. Cambridge, MA: Harvard University Press, 1988.

Lambrecht, Bill. *Dinner at the New Gene Café: How Genetic Engineering Is Changing What We Eat, How We Live, and the Global Politics of Food*. New York: St. Martin's Press, 2001.

Laxness, Halldór. J. A. Thompson, trans. *Independent People*. New York: Vintage, 1997 (Icelandic original, 1946).

Le, Stephen. *100 Million Years of Food*. New York: Picador, 2016.

Leslie, Eliza. *Miss Leslie's Complete Directions for Cookery*. Philadelphia: E. L. Carey and A. Hart, 1837.

Lysaght, Patricia, ed. *Milk and Milk Products from Medieval to Modern Time: Proceedings of the Ninth International Conference on Ethnological Food Research*. Edinburgh: Canongate Press, 1994.

MacDonogh, Giles. *A Palate in Revolution: Grimod de La Reynière and the Almanach des Gourmands*. London: Robin Clark, 1987.

Markham, Gervase. Michael R., Best, ed. *The English Housewife* (1615). Montreal: McGill–Queen's Press, 1986.

Marshall, A. B. *Ices Plain & Fancy*. New York: Metropolitan Museum of Art, 1976.

———. *Mrs. A. B. Marshall's Cookery Book*. London: Ward, Lock & Co., 1887.

Mason, Laura, and Catherine Brown. *The Taste of Britain*. London: Harper Press, 2006.

Masters, Thomas. *The Ice Book: A History of Everything Connected with Ice, with Recipes*. London: Simpkin, Marshall & Company, 1844.

May, Robert. *The Accomplisht Cook*. London: Bear and Star in St. Paul's Churchyard, 1685.

McCleary, George Frederick. *The Municipalization of the Milk Supply*. London: Fabian Municipal Program, 2nd series, no. 1, 1902.

Mendelson, Anne. *Milk: The Surprising Story of Milk through the Ages*. New York: Alfred A. Knopf, 2008.

Milham, Mary Ella, trans. *Platina: On Right Pleasure and Good Health*. Tempe, AZ: Medieval and Renaissance Text and Studies, 1998.

Montagne, Prosper. *Larousse Gastronomique*. Paris: Larousse, 1938.

Pant, Pushpesh. *India Cookbook*. London: Phaidon, 2010.

Peachey, Stuart. *Civil War and Salt Fish: Military and Civilian Diet in the C17*. Essex, England: Partizan Press, 1988.

Pidathala, Archana. *Five Morsels of Love*. Hyderabad: Archana Pidathala, 2016.

Pliny the Elder. John F. Healy, trans. *Natural History: A Selection*. New York: Penguin, 1991.

Polo, Marco. Ronald Latham, trans. *The Travels*. New York: Penguin, 1958.

Porterfield, James D. *Dining by Rail*. New York: St. Martin's/Griffin, 1993.

Powell, Marilyn. *Ice Cream: The Delicious History*. Woodstock, NY: Overlook Press, 2005.

Prasada, Neha, and Ashima Narain. *Dining with the Maharajahs: A Thousand Years of Culinary Tradition*. New Delhi: Roli Books, undated.

Prato, C. *Manuale di Cucina*. Milan: Anonima Libraria Italiano, 1923.

Prudhomme, Paul. *Chef Paul Prudhomme's Louisiana Kitchen*. New York: William Morrow, 1984.

Quinzio, Jeri. *Of Sugar and Snow: A History of Ice Cream Making*. Berkeley: University of California Press, 2009.

Raffald, Elizabeth. *The Experienced English Housekeeper: For the Use and Ease of Ladies, Housekeepers, Cooks etc*. London: 1782.

Ragnarsdóttir, Thorgerdur. *Skyr: For 1000 Years*. Reykjavik: MensMentis ehf, 2016.

Randolph, Mary. *The Virginia Housewife: or Methodical Cook*. Baltimore: Plakitt & Cugell, 1824.

Ranhofer, Charles. *A Complete Treatise of Analytical and Practical Studies of the Culinary Art*. New York: R. Ranhofer, 1893.

Rawlings, Marjorie Kinnan. *Cross Creek Cookery*. New York: C. Scribner's Sons, 1942.

Reboul, J.-B. *La Cuisinière Provençale*. Marseille: Tacussel, 1897.

Rorer, Sarah Tyson Heston. *Fifteen New Ways for Oysters*. Philadelphia: Arnold and Company, 1894.

———. *Ice Creams, Water Ices, Frozen Puddings, Together with Refreshments for All Social Affairs*. Philadelphia: Arnold and Company, 1913.

Richardson, Tim. *Sweets: A History of Candy*. New York: Bloomsbury, 2002.

Riley, Gillian. *The Dutch Table*. San Francisco: Pomegranate Artbook, 1994.

Rodinson, Maxime, A. J. Arberry, and Charles Perry. *Medieval Arab Cookery*. Devon, England: Prospect Books, 2001.

Rose, Peter G. *The Sensible Cook: Dutch Foodways in the Old and New World*. Syracuse, NY: Syracuse University Press, 1989.

Rosenau, M. J. *The Milk Question*. London: Constable & Co., 1913.

Sand, George. *Scènes Gourmandes*. Paris: Librio, 1999.

Schama, Simon. *The Embarrassment of Riches: An Interpretation of Dutch Culture in the Golden Age*. London: Harper Perennial: 2004.

Scully, Eleanor, and Terence Scully. *Early French Cookery: Sources, History, Original Recipes and Modern Adaptions*. Ann Arbor: University of Michigan Press, 1995.

Scully, Terence, ed. *Chiquart's "On Cookery:" A Fifteenth-Century Savoyard Culinary Treatise*. New York: Peter Lang, 1986.

———. *The Viandier of Taillevent*. Ottawa: University of Ottawa Press, 1988.

Seely, Lida. *Mrs. Seely's Cook Book: A Manual of French and American Cookery*. New York: Macmillan, 1902.

Selitzer, Ralph. *The Dairy Industry in America*. New York: Dairyfield, 1976.

Sereni, Clara. Giovanna Miceli Jeffries and Susan Briziarelli, trans. *Keeping House: A Novel in Recipes*. Albany: State University of New York Press, 2005.

Simmons, Amelia. *American Cookery*. Albany: George R. & George Webster, 1796.

Smith, Eliza. *The Compleat Housewife*. London: 1758.

Smith-Howard, Kendra. *Pure and Modern Milk: An Environmental History Since 1900*. Oxford: Oxford University Press, 2014.

Spargo, John. *The Common Sense of the Milk Question*. New York: Macmillan, 1908.

Spaulding, Lily May, and John Spaulding, eds. *Civil War Recipes: Receipts from the Pages of Godey's Lady's Book*. Lexington: University Press of Kentucky, 1999.

Spencer, Colin. *British Food: An Extraordinary Thousand Years of History*. New York: Columbia University Press, 2003.

Stout, Margaret B. *The Shetland Cookery Book*. Lerwick: T. & J. Manson, 1965.

Straus, Nathan. *Disease in Milk: The Remedy, Pasteurization*. New York, 1913.

Sullivan, Caroline. *The Jamaican Cookery Book*. Kingston: Aston W. Gardner & Co., 1893.

Terrail, Claude. *Ma Tour d'Argent*. Paris: Marabout, 1975.

Thibaut-Comelade, Éliane. *La Table Medieval des Catalans*. Montpellier: Les Presses du Languedoc, 1995.

Thornton, P. *The Southern Gardener and Receipt Book*. Newark, NJ: by the author, 1845.

Toomre, Joyce, ed. *Elena Molokhovets' "A Gift to Young Housewives."* Bloomington: Indiana University Press, 1992.

Thorsson, Örnólfur. *The Sagas of the Icelanders*. New York: Penguin, 2001.

Toklas, Alice B. *The Alice B. Toklas Cook Book*. New York: Harper & Brothers, 1954.

Tschirky, Oscar. *The Cookbook by "Oscar" of the Waldorf.* New York: Werner Company, 1896.

Twamley, Josiah. *Dairy Exemplified, or the Business of Cheese Making.* London: Josiah Twamley, 1784.

Valenze, Deborah. *Milk: A Local and Global History.* New Haven: Yale University Press, 2011.

Vallery-Radot, Rene. *Louis Pasteur.* New York: Alfred A. Knopf, 1958.

Van Ingen, Philip, and Paul Emmons Taylor, eds. *Infant Mortality and Milk Stations.* New York: New York Milk Committee, 1912.

Vehling, Joseph Dommers, ed. and trans. *Apicius: Cookery and Dining in Imperial Rome.* New York: Dover, 1977.

Verrall, William. *Cookery Book.* Lewes, East Sussex: Southover Press, 1988. First published 1759.

Walker, Harlan, ed. *Milk: Beyond the Dairy: Proceedings of the Oxford Symposium on Food and Cookery, 1999.* Totnes, Devon: Prospect Books, 2000.

Wilkins, John, David Harvey, and Mike Dobson, eds. *Food in Antiquity.* Exeter: University of Exeter Press, 1995.

William of Rubruck. William Woodville Rockhill, trans. *The Journey of William of Rubruck to the Eastern Parts of the World, 1253–55.* London: Hakluyt Society, 1940.

Wilson, Avice R. *Cocklebury: A Farming Area and Its People in the Vale of Wiltshire.* Chichester, Sussex: Phillimore, 1983.

———. *Forgotten Harvest: The Story of Cheesemaking in Wiltshire.* Wiltshire: Avice R. Wilson, 1995.

ARTICLES

American Academy of Pediatrics. "Breastfeeding and the Use of Human Milk." *Pediatrics* 115, no. 2 (February 1, 2005).

Biotechnical Information Series, "Bovine Somatotropin (bST)," Iowa State University, December 1993

Couch, James Fitton. "The Toxic Constituent of Richweed or White Snakeroot (*Eupatorium urticaefolium*)," *Journal of Agricultural Research* 35, no. 6 (September 15, 1927).

Hakim, Danny. "Doubts about a Promised Bounty." *New York Times,* October 30, 2016.

McCracken, Robert D. "Lactase Deficiency: An Example of Dietary Evolution." *Current Anthropology* 12, no. 4–5: 479–517.

Noble, Josh. "Asia's Bankers Milk China Thirst for Dairy Products." *Financial Times*, November 12, 2013.

Poo, Mu-chou. "Liquids in Temple Ritual." UCLA Encyclopedia of Egyptology, September 25, 2010.

Tavernise, Sabrina, "F.D.A. Restricts Antibiotics Use for Livestock." *New York Times*, December 11, 2013.

RECIPE INDEX

PANCAKES, PORRIDGE AND SCONES

CHEESE DISHES

SOUPS

SAUCES

BUTTER

SAVORY MILK DISHES

YOGURT DISHES

INDEX

Note: Italic page numbers refer to illustrations.

A NOTE ON THE AUTHOR

MARK KURLANSKY is the *New York Times* bestselling author
of *Cod, Salt, Paper, Havana, The Basque History of the World, 1968,* and
The Big Oyster, among other titles. He has received the Dayton Literary
Peace Prize, *Bon Appétit's* Food Writer of the Year Award, the James
Beard Award, and the Glenfiddich Award. He lives in New York City.

www.markkurlansky.com
Twitter: @codlansky